AF361687

Metabolomics, Oxidative, and Nitrosative Stress in the Perinatal Period

Metabolomics, Oxidative, and Nitrosative Stress in the Perinatal Period

Editors

Julia Kuligowski
Máximo Vento

MDPI • Basel • Beijing • Wuhan • Barcelona • Belgrade • Manchester • Tokyo • Cluj • Tianjin

Editors
Julia Kuligowski
Neonatal Research Group
Health Research Institute La
Fe
Valencia
Spain

Máximo Vento
Neonatal Research Group
Health Research Institute La
Fe
Valencia
Spain

Editorial Office
MDPI
St. Alban-Anlage 66
4052 Basel, Switzerland

This is a reprint of articles from the Special Issue published online in the open access journal *Antioxidants* (ISSN 2076-3921) (available at: www.mdpi.com/journal/antioxidants/special_issues/ Metabolomics_Perinatal_Period).

For citation purposes, cite each article independently as indicated on the article page online and as indicated below:

LastName, A.A.; LastName, B.B.; LastName, C.C. Article Title. *Journal Name* **Year**, *Volume Number*, Page Range.

ISBN 978-3-0365-5052-7 (Hbk)
ISBN 978-3-0365-5051-0 (PDF)

Contents

About the Editors

Julia Kuligowski

Julia Kuligowski is a senior researcher at the Neonatal Research Group of the Health Research Institute La Fe (Valencia, Spain) and Short-Term Scientific Mission Coordinator of the Pan-European Network in Lipidomics and EpiLipidomics (CA19105). She is an expert on metabolomics and lipidomics, as well as the development of quantitative bioanalytical methods. For over ten years, she has been working on liquid and gas chromatography coupled to mass spectrometry, as well as vibrational spectroscopy, employed as research tools focusing on the analysis biofluids from newborns as well as from human milk. Currently, she is leading research projects supported by national and international funding sources on the impact of inhaled nitric oxide on oxidative and nitrosative stress, as well as the benefits of human milk on preterm infant health. She is involved in the Metabolomics Quality Assurance and Quality Control Consortium (mQACC) and has published over 120 papers in peer-reviewed, international journals, as well as over 10 book chapters.

Máximo Vento

Máximo Vento MD PhD, Professor of Pediatrics, is presently an Emeritus Investigator and Principal Investigator of the Neonatal Research Group at the Health Research Institute La Fe (Valencia, Spain) and Chairman of the European Board of Neonatology under the auspices of the European Society for Paediatric Research. His main areas of interest comprise the physiology and pathophysiology of the fetal to neonatal transition and postnatal stabilization, oxygen metabolism and toxicity, infection, nutrition, and respiratory conditions from an experimental and clinical perspective in the neonatal, in both term and preterm infants. He has pioneered the use of modern analytical techniques for developing biomarkers to assess and predict outcomes in the most relevant conditions during the perinatal period. As a result, he has published more than 400 peer-reviewed articles in relevant international journals, as well as 50 book chapters.

Preface to "Metabolomics, Oxidative, and Nitrosative Stress in the Perinatal Period"

Studies focusing on the perinatal period face unique challenges, yet research in this area is extremely important, as this period of life is highly delicate and adverse events might have a long-lasting impact. With the advent of powerful high-resolution and high-throughput analytical methods, researchers have started to successfully develop and implement novel approaches in this area. New insights have great potential to be translated into novel diagnostic tools, as well as alternative preventive and treatment approaches. This book collects a series of timely review and original research articles focusing on metabolomic, oxidative, and nitrosative stress in the perinatal period.

We would like to thank all involved authors for their high-quality contributions and their commitment to the publication of this work and hope that this book will be a useful resource for students, scientists, and doctors working in this specific area of application.

Julia Kuligowski and Máximo Vento
Editors

antioxidants

MDPI

Editorial

Metabolomics, Oxidative, and Nitrosative Stress in the Perinatal Period

Julia Kuligowski [1,*] and Máximo Vento [1,2]

[1] Neonatal Research Group, Health Research Institute La Fe (IISLaFe), 46026 Valencia, Spain; maximo.vento@uv.es
[2] Division of Neonatology, University and Polytechnic Hospital La Fe (HULAFE), 46026 Valencia, Spain
* Correspondence: julia.kuligowski@uv.es; Tel.: +34-961-246-661

Citation: Kuligowski, J.; Vento, M. Metabolomics, Oxidative, and Nitrosative Stress in the Perinatal Period. *Antioxidants* **2022**, *11*, 1357. https://doi.org/10.3390/antiox11071357

Received: 8 July 2022
Accepted: 8 July 2022
Published: 12 July 2022

Publisher's Note: MDPI stays neutral with regard to jurisdictional claims in published maps and institutional affiliations.

The perinatal period is extremely sensitive to external stimuli, and events that may disturb the equilibrium within the mother–infant dyad might have a substantial short- and long-term impact on the infant's health and development. Lacking oxygen in the perinatal period can result in bioenergetic failure and cell death, with consequences for infant health and survival. While the administration of supplemental concentrations of oxygen can be lifesaving, it can also disrupt growth and development, impair bioenergetic function, and induce inflammation. Oxidative stress is of key importance in the pathophysiology of several diseases, affecting term as well as preterm infants. Professor OD Saugstad referred to these effects as the "Oxygen radical diseases of Neonatology" [1]. This important topic is addressed in this Special Issue in a review article which describes the historical perspective and current trends in the use of supplemental oxygen in newborns, covering unique features of newborn redox physiology and antioxidant defenses, the history of therapeutic oxygen use in this population and its role in disease, clinical trends in the use of therapeutic oxygen and mitigation of neonatal oxidative injury [2].

Sex differences in the susceptibility to oxidative-stress-related complications of prematurity are a widely accepted concept. The foundation of this theory is elucidated in a systematic review and meta-analysis of cohort studies exploring the association between sex and complications of prematurity, showing that male preterm male infants have higher clinical instability and greater need for invasive interventions than female preterm infants [3]. This leads to a male disadvantage in mortality and explains differences in short-term complications of prematurity.

Human milk is a globally recognized gold standard for infant nutrition, and is especially important for the healthy growth and development of preterm infants. Furthermore, the antioxidant properties of human milk mitigate the consequences of excessive oxidative damage. When the mother's own milk is unavailable, pasteurized donor milk is the best alternative. Although pasteurization is necessary for safety reasons, it may affect the concentration and activity of biological factors, including antioxidants. This Special Issue includes a review article describing the effect of different pasteurization methods on antioxidant properties of human milk that aims to provide evidence to guide donor milk banks in choosing the best pasteurization method from an antioxidant perspective [4].

In recent years, and with the advent of powerful high resolution and high-throughput analytical methods, researchers have begun to successfully develop and implement novel metabolomics approaches for diagnostics in perinatal asphyxia. It is important to design non-invasive tools tailored to the characteristics of newborns, and saliva analysis has unexplored potential. This Special Issue includes a literature review providing an overview of metabolomics studies of oxidative stress in perinatal asphyxia, particularly seeking studies analyzing non-invasively collected biofluids such as saliva [5]. While changes in oxidative stress-related salivary metabolites have been reported in adults, the utility of this approach in perinatal asphyxia has yet to be explored. Experimental and clinical studies indicate that, in addition to antioxidant enzymes, succinate and hypoxanthine, acylcarnitines may also have discriminatory diagnostic and prognostic properties in perinatal asphyxia. Accumulating evidence of discriminatory metabolic patterns in perinatal asphyxia may be useful to develop point-of-care methods for measuring salivary oxidative stress metabolites relevant in perinatal asphyxia.

The versatility of metabolomics and the use of this methodology for a sensitive detection of transient changes in longitudinal studies has been demonstrated in a study involving newborns with transposition of the great arteries (TGA), a common cyanotic congenital heart disease [6]. Parallel circulations that result in impaired cerebral oxygen delivery in utero may lead to brain damage and long-term neurodevelopmental delay. To mix deoxygenated and oxygenated blood at the atrial level, balloon atrial septostomy (BAS) is frequently employed, causing a sudden increase in arterial blood oxygenation and oxidative stress. Changes in oxygen saturation as well as metabolic profiles of plasma samples from nine newborn infants suffering from TGA before and until 48 h after undergoing BAS were recorded. The plasma metabolome changed clearly over time, and alterations of four metabolic pathways, including the pentose phosphate pathway, were linked to changes in the cerebral tissue oxygen extraction. On the contrary, no changes in levels of lipid peroxidation biomarkers were observed. These findings suggest that metabolic adaptations buffer the free radical burst triggered by re-oxygenation, thereby avoiding structural damage at the macromolecular level.

Neonatal encephalopathy is one of the main causes of morbidity and mortality in term infants, and there is evidence that oxidative damage plays an important role in the pathophysiology of hypoxic–ischemic brain injury. Recently, several studies have been conducted with the aim of providing adjacent neuroprotective and antioxidant therapeutic options complementary to hypothermia treatment, animal [7–9] and clinical [10] studies with different compounds. Kynurenic acid significantly reduced reactive oxygen species and antioxidant enzyme activity and enhanced GSH levels and hypoxic–ischemic conditions in a rat model [7]. Only the highest concentration of kynurenic acid showed neuroprotection when applied 6 h after hypoxia–ischemia, and results indicated the induction of neuroprotection at the reactive oxygen species formation stage. In a different study, the capability of postnatal allopurinol administration in combination with hypothermia treatment to reduce oxidative stress biomarkers was assessed in a rat model of hypoxic–ischemic encephalopathy, showing that the administration of allopurinol, hypothermia, and the combination treatment protects the brain against oxidative damage [8]. Furthermore, in neonatal hypoxia ischemia complicated by perinatal infection, therapeutic hypothermia does not improve outcomes due to pre-existing oxidative stress and neuroinflammation, which shorten the therapeutic window. Hence, the definition and targeting of central nervous system metabolomic changes immediately after endotoxin-sensitized hypoxia–ischemia (LPS-HI) has been studied in a rat model with the aim of discovering neuroprotection strategies that are effective post-injury [9]. Despite hypothermia treatment, LPS-HI acutely depleted reduced glutathione, indicating overwhelming oxidative stress. However, the combination of the administration of N-acetylcysteine and vitamin D (NVD) with hypothermia rapidly improved cellular redox status after LPS-HI, potentially through the inhibition of important secondary injury cascades, allowing more time for hypothermic neuroprotection. The promising results obtained in this experimental study were further validated in a clinical trial that aimed to translate these FDA-approved drugs to HIE neonates [10]. NVD was well tolerated and 24 treated HIE infants had no evidence of cerebral palsy, autism, or cognitive delay at 24–48 months, confirming that low, safe doses of NVD in HIE neonates decrease oxidative stress, improve central nervous system energetics, and are associated with favorable long-term developmental outcomes.

In summary, this Special Issue demonstrates the potential of the use of metabolomics and the assessment of oxidative stress for gaining insight in a variety of relevant aspects related to the health and disease of newborns, as well as new therapeutic approximations.

Funding: This research was funded by Instituto de Salud Carlos III and co-funded by the European Union (grant number CP16/00034; and project number PI20/00964); the Spanish Maternal, Neonatal and Developmental Network SAMID RETICS (grant number RD16/0022/0001) funded by the PN2018-2021 (Spain), ISCIII-Sub-Directorate General for Research Assessment and Promotion and the European Regional Development Fund (FEDER).

Conflicts of Interest: The authors declare no conflict of interest.

References

1. Saugstad, O.D. Oxygen radical disease in neonatology. *Semin. Neonatol.* **1998**, *3*, 229–238. [CrossRef]
2. Mathias, M.; Chang, J.; Perez, M.; Saugstad, O. Supplemental Oxygen in the Newborn: Historical Perspective and Current Trends. *Antioxidants* **2021**, *10*, 1879. [CrossRef] [PubMed]
3. van Westering-Kroon, E.; Huizing, M.J.; Villamor-Martínez, E.; Villamor, E. Male Disadvantage in Oxidative Stress-Associated Complications of Prematurity: A Systematic Review, Meta-Analysis and Meta-Regression. *Antioxidants* **2021**, *10*, 1490. [CrossRef] [PubMed]
4. Juncker, H.G.; Ruhé, E.J.M.; Burchell, G.L.; van den Akker, C.H.P.; Korosi, A.; van Goudoever, J.B.; van Keulen, B.J. The Effect of Pasteurization on the Antioxidant Properties of Human Milk: A Literature Review. *Antioxidants* **2021**, *10*, 1737. [CrossRef] [PubMed]
5. Solevåg, A.L.; Zykova, S.N.; Thorsby, P.M.; Schmölzer, G.M. Metabolomics to Diagnose Oxidative Stress in Perinatal Asphyxia: Towards a Non-Invasive Approach. *Antioxidants* **2021**, *10*, 1753. [CrossRef] [PubMed]
6. Piñeiro-Ramos, J.D.; Rahkonen, O.; Korpioja, V.; Quintás, G.; Pihkala, J.; Pitkänen-Argillander, O.; Rautiainen, P.; Andersson, S.; Kuligowski, J.; Vento, M. A Reductive Metabolic Switch Protects Infants with Transposition of Great Arteries Undergoing Atrial Septostomy against Oxidative Stress. *Antioxidants* **2021**, *10*, 1502. [CrossRef] [PubMed]
7. Bratek-Gerej, E.; Ziembowicz, A.; Godlewski, J.; Salinska, E. The Mechanism of the Neuroprotective Effect of Kynurenic Acid in the Experimental Model of Neonatal Hypoxia–Ischemia: The Link to Oxidative Stress. *Antioxidants* **2021**, *10*, 1775. [CrossRef]
8. Durán Fernández-Feijóo, C.; Rodríguez-Fanjul, J.; Lopez-Abat, M.; Hadley, S.; Cavia-Saiz, M.; Muñiz, P.; Arnaez, J.; Fernández-Lorenzo, J.R.; Camprubí Camprubí, M. Effects of Hypothermia and Allopurinol on Oxidative Status in a Rat Model of Hypoxic Ischemic Encephalopathy. *Antioxidants* **2021**, *10*, 1523. [CrossRef] [PubMed]
9. Adams, L.E.; Moss, H.G.; Lowe, D.W.; Brown, T.; Wiest, D.B.; Hollis, B.W.; Singh, I.; Jenkins, D.D. NAC and Vitamin D Restore CNS Glutathione in Endotoxin-Sensitized Neonatal Hypoxic-Ischemic Rats. *Antioxidants* **2021**, *10*, 489. [CrossRef] [PubMed]
10. Jenkins, D.D.; Moss, H.G.; Brown, T.R.; Yazdani, M.; Thayyil, S.; Montaldo, P.; Vento, M.; Kuligowski, J.; Wagner, C.; Hollis, B.W.; et al. NAC and Vitamin D Improve CNS and Plasma Oxidative Stress in Neonatal HIE and Are Associated with Favorable Long-Term Outcomes. *Antioxidants* **2021**, *10*, 1344. [CrossRef] [PubMed]

Short Biography of Authors

Julia Kuligowski is a senior researcher at the Neonatal Research Group of the Health Research Institute La Fe (Valencia, Spain) and Short-Term Scientific Mission Coordinator of the Pan-European Network in Lipidomics and EpiLipidomics (CA19105). She is an expert on metabolomics and lipidomics as well as the development of quantitative bioanalytical methods. For over ten years, she has been working on liquid and gas chromatography coupled to mass spectrometry as well as vibrational spectroscopy employed as research tools focusing on the analysis biofluids from the newborn as well as human milk. Currently, she is leading research projects supported by national and international funding sources on the impact of inhaled nitric oxide on oxidative and nitrosative stress, as well as the benefits of human milk on preterm infant health. She is involved in the Metabolomics Quality Assurance and Quality Control Consortium (mQACC) and has published over 120 papers in peer-reviewed, international journals as well as over 10 book chapters.

Maximo Vento Professor of Pediatrics is at present Emeritus Investigator and Principal Investigator of the Neonatal Research Group at the Health Research Institute La Fe (Valencia, Spain), and Chairman of the European Board of Neonatology under the auspices of the European Society for Paediatric Research. His main areas of interest comprise the physiology and pathophysiology of the fetal to neonatal transition and postnatal stabilization, oxygen metabolism and toxicity, infection, nutrition, and respiratory conditions from an experimental and clinical perspective in the neonatal both in term and preterm infants. He has pioneered the use of modern analytical techniques for developing biomarkers to assess and predict outcomes in the most relevant conditions during the perinatal period. As a result, he has published more than 400 peer-reviewed articles in relevant international journals, and 50 book chapters.

Review

Supplemental Oxygen in the Newborn: Historical Perspective and Current Trends

Maxwell Mathias [1],* , Jill Chang [2,3], Marta Perez [2,3] and Ola Saugstad [2,4]

[1] Center for Pregnancy and Newborn Research, Department of Pediatrics, Section of Neonatal-Perinatal Medicine, University of Oklahoma Health Sciences Center, Oklahoma City, OK 73104, USA

[2] Division of Neonatology, Department of Pediatrics, Northwestern University Feinberg School of Medicine, Chicago, IL 60611, USA; jchang@luriechildrens.org (J.C.); mtperez@luriechildrens.org (M.P.); o.d.saugstad@medisin.uio.no (O.S.)

[3] Ann & Robert H. Lurie Children's Hospital of Chicago, Chicago, IL 60611, USA

[4] Department of Pediatric Research, University of Oslo, N-0424 Oslo, Norway

* Correspondence: maxwell-mathias@ouhsc.edu; Tel.: +1-405-271-5215; Fax: +1-405-271-1236

Abstract: Oxygen is the final electron acceptor in aerobic respiration, and a lack of oxygen can result in bioenergetic failure and cell death. Thus, administration of supplemental concentrations of oxygen to overcome barriers to tissue oxygen delivery (e.g., heart failure, lung disease, ischemia), can rescue dying cells where cellular oxygen content is low. However, the balance of oxygen delivery and oxygen consumption relies on tightly controlled oxygen gradients and compartmentalized redox potential. While therapeutic oxygen delivery can be life-saving, it can disrupt growth and development, impair bioenergetic function, and induce inflammation. Newborns, and premature newborns especially, have features that confer particular susceptibility to hyperoxic injury due to oxidative stress. In this review, we will describe the unique features of newborn redox physiology and antioxidant defenses, the history of therapeutic oxygen use in this population and its role in disease, and clinical trends in the use of therapeutic oxygen and mitigation of neonatal oxidative injury.

Keywords: hyperoxia; prematurity; bronchopulmonary dysplasia; retinopathy of prematurity

Citation: Mathias, M.; Chang, J.; Perez, M.; Saugstad, O. Supplemental Oxygen in the Newborn: Historical Perspective and Current Trends. *Antioxidants* **2021**, *10*, 1879. https://doi.org/10.3390/antiox10121879

Academic Editors: Julia Kuligowski and Máximo Vento

Received: 14 October 2021
Accepted: 23 November 2021
Published: 25 November 2021

Publisher's Note: MDPI stays neutral with regard to jurisdictional claims in published maps and institutional affiliations.

1. Introduction

1.1. Hyperoxia Is Damaging to Developing Organ Systems

Mammals gestate at lower oxygen tension than their postnatal environment, and birth (even in the absence of supplemental oxygen) constitutes an increase in oxygen exposure [1]. In addition, rapid post-natal growth and development requires significant oxygen delivery and consumption per gram of tissue in newborns, and energy sources are more rapidly depleted than in adults [2,3]. Alterations in mitochondrial oxygen concentration can induce apoptotic cell signaling pathways through mitochondrial proton leak [4], while non-mitochondrial derived reactive oxygen and nitrogen intermediates (ROI) play an important role in intra- and inter-cellular growth factor signaling (see [5] for review). Lastly, enzymatic and non-enzymatic antioxidant systems in newborns have decreased capacity to sequester ROI [6]. Premature exposure to ambient air (fraction of inspired oxygen or FiO_2 = 21%) and oxygen exposure above ambient air (FiO_2 > 21%) disrupts these processes.

The preterm and term newborn occupies a unique position as an organism with high tissue oxygen demand, growing and developing organ systems, and susceptibility to oxidative injury. Excess oxygen is the substrate for free radical formation by several enzymes, such as the xanthine oxidase and NADPH-oxidase families [7]. Free radicals induce membrane disruption and activation of inflammatory pathways through lipid peroxidation, affecting multiple organ systems. Elevated lipid peroxidation products have been found in plasma samples of infants with bronchopulmonary dysplasia (BPD) compared to those without [8]. In the lungs, hyperoxia exposure induces alveolar simplification and vascular remodeling, which is associated with disruption in the electron transport chain [9,10]. Hyperoxia exposure is also associated with injury and altered physiology in the developing intestines and retinopathy of prematurity (ROP), a developmental eye disease involving aberrant growth of the retinal vasculature that can result in blindness [11–13]. Lastly, neonatal hyperoxia alters cerebral blood flow, and neuronal apoptosis and inflammation have been shown after neonatal hyperoxia exposure [14–16].

1.2. Full Term and Preterm Newborns Have Decreased Antioxidant Capacity

Many important vitamin and mineral stores are delivered to the fetus during the third trimester of pregnancy; as a result, preterm infants have decreased levels of these important cofactors relative to term infants [17]. Plasma vitamin C levels are lower in preterm infants relative to term infants [18]. In a prospective study of vitamin D supplementation in 100 preterm infants, two-thirds had biochemical vitamin D deficiency [19]. In addition, pulmonary levels of glutathione peroxidase 1, cytosolic superoxide dismutase (SOD), mitochondrial SOD, and extracellular SOD are significantly lower in newborn mice than those at 3 and 7 weeks of age in mice. Along with decreased antioxidant expression, newborn mice exhibit increased mitochondrial oxidative stress compared to adults during hyperoxia exposure [6]. In a cohort of preterm infants with and without BPD, plasma vitamin E levels were below term infant levels, and were lower in infants with BPD than those without [20].

Given the decreased levels of antioxidant enzymes and vitamins in preterm infants, several clinical trials have been conducted to evaluate whether supplementation with either intact antioxidant enzyme or vitamin cofactors might decrease morbidity and mortality in this population. Trials of administration of exogenous SOD in the clinical setting were initially promising [21–23]; however, meta-analyses and long term follow-ups have shown no benefit over placebo [24]. Similarly, supplementation with the antioxidant cofactor vitamins A, C, and E have been studied, along with the glutathione precursor N-acetylcysteine (NAC). Vitamin A has been the subject of extensive research with mixed results, including the largest randomized controlled trial to date entitled NeoVitaA, which has not yet published its results [25–27]. A Cochrane meta-analysis in 2016 found marginal benefit of intramuscular administration of vitamin A to prevent death or BPD (RR 0.93 [0.88, 0.99]); however, intramuscular formulation is not readily available and repeat intramuscular injections pose a practical challenge in very low birth weight (<1500 g) infants [28]. Vitamin C, E, and NAC have also not shown clear benefit, and supplementation beyond nutritional needs is not recommended [29–34].

1.3. ROI Are Essential Components of Cell Signaling Pathways in Development

Despite the association of neonatal disease with ROI byproducts (e.g., lipid peroxides), the growth and development of organs and tissues relies on the production of ROI. Numerous growth factor receptors are tyrosine kinases (RTKs), whose function depends on local hydrogen peroxide derived from ROI. Peroxide oxidizes protein phosphatases (PTPs) by altering the 3-D structure and allowing RTK autophosphorylation and signal transduction [35,36]. Suppression of hydrogen peroxide with exogenous catalase will reduce tyrosine kinase phosphorylation and decrease growth factor signaling, which has been shown to reduce vascular endothelial cell migration [37,38]. In addition, increases in antioxidant capacity can impair growth and development. Inhibition of nuclear factor erythroid 2-related factor 2 (NRF2), an important transcription factor in the production of antioxidants in response to oxidative stress, increased fetal growth in a mouse model of pregnancy-associated hypertension, suggesting an important role for ROI in fetal growth [39]. A clearer picture of the complex interplay between postnatal oxygen exposure, antioxidant expression and function, and development and growth signaling is needed.

2. Supplemental Oxygen in the Initial Resuscitation of Newborns

2.1. Historical Aspects

While the first documented use of oxygen in infants was in 1780, just few years after its discovery, it was not until 1928 that a description of using oxygen in newborn resuscitation was published [40]. Oxygen therapy was subsequently introduced into routine newborn care in the 1930s and early 1940s. This was followed in relatively short order by the first published description of retrolental fibroplasia (RLF), a mysterious eye disease found in a small group of infants born prematurely, now known as ROP [41]. The progressive nature of this disorder was confirmed in 1948 by a husband-and-wife opthalmologist team working at Johns Hopkins University, who described development of RLF changes in a cohort of premature infants [42]. It was not until the 1950s that oxygen was identified as the culprit behind RLF, the leading cause of blindness in children of preschool age at that time [43–45]. This discovery was followed by decades of avoidance of oxygen that likely led to decreased survival of the most premature and sickest infants [46]. With the advent of oxygen saturation monitoring in the 1980s, however, oxygen therapy could be better targeted, resulting in improved survival and an associated increase in ROP once again [47].

2.2. Supplemental Oxygen in Term Infants

By the 1960s, the use of 100% oxygen in the delivery room was viewed as the standard of care and the most sensible approach to resuscitation of asphyxiated infants [48]. The International Liaison Committee on Resuscitation (ILCOR), first formed in 1992 to provide a forum for major resuscitation organizations in the industrialized world, recommended the use of 100% oxygen in its first set of newborn guidelines [49]. There were published opinions against such an approach as early as 1980, based on emerging clinical and animal data that implicated delivery room hyperoxia in subsequent oxidative stress generation [7,50]. This included evidence of increased hypoxanthine concentrations in the cerebral cortex of hypoxemic piglets randomized to resuscitation with 100% FiO_2 compared to 21% FiO_2, suggesting the presence of more severe energy metabolism deficits in the hyperoxia-exposed group. Animal models also increasingly supported the use of 21% FiO_2 during resuscitation and demonstrated that it is as effective as 100% FiO_2 in the resuscitation of hypoxemic piglets, with similar improvements in vital signs, base deficit, and plasma hypoxanthine [51,52]. Furthermore, studies in human neonates demonstrated that term newborns can be adequately resuscitated with 21% FiO_2, with lower mortality in infants exposed to 21% vs. 100% FiO_2 [53], with follow-up studies showing similar long-term neurodevelopmental outcomes [54]. Additional studies demonstrated that resuscitation with 100% FiO_2 can be harmful [55,56]. In term neonates with perinatal asphyxia, resuscitation with 21% FiO_2 resulted in more rapid improvements in Agpar scores and faster onset of spontaneous respiration, while infants treated with high FiO_2 demonstrated evidence of increased oxidative stress at one month of age, with lower reduced-to-oxidized glutathione ratios and higher erythrocyte antioxidant enzyme activity [55]. Finally, a meta-analysis also contributed to the growing consensus; in term and late-preterm neonates in need of resuscitation at birth, 21% FiO_2 restores heart rate and spontaneous respiration as effectively as 100% FiO_2, while also being associated with significant reductions in neonatal mortality [57]. This accumulating evidence resulted in revisions to neonatal resuscitation guidelines that now recommend starting with 21% FiO_2 and caution against 100% FiO_2 in near-term and in-term infants [58,59].

2.3. Supplemental Oxygen in Preterm Infants

Evidence for the optimal oxygen concentration in the resuscitation of preterm infants is less conclusive. The available data suggest that many preterm infants need an FiO_2 somewhere between 21% and 100% during resuscitation. A recently published meta-analysis found conflicting data on the impact of initiating resuscitation with lower (<40%) vs. higher ($\geq$40%) FiO_2 in preterm infants, with 9 out of 10 studies finding no differences in mortality [60]. Other publications, however, found an association between initiation of resuscitation with lower FiO_2 and increased risk of death in immature preterm infants in an international cohort [61] and a Canadian cohort [62]. However, a follow-up study of the Canadian cohort at 18 to 21 months corrected age (gestational age at birth plus absolute age) found no differences in the composite outcome of death or neurodevelopmental impairment and an increased risk of severe neurodevelopmental impairment in infants who received 100% FiO_2 versus those receiving 21% FiO_2 [63]. However, preterm infants not reaching an oxygen saturation (SpO_2) of 80% within the first 5 min of life have been shown to have higher mortality and risk of severe IVH [64,65]. In addition, the combination of bradycardia and hypoxemia in the first minutes of life wasn't increases mortality. Although the optimal initial FiO_2 is not fully known in these infants, it is still recommended to target a SpO_2 of 80–85% within the first 5 min of life, as well as to avoid bradycardia (heart rate < 100 bpm).

3. Supplemental Oxygen in the Neonatal ICU

3.1. Historical Perspective

Beyond the delivery room, exposure to hyperoxia and oxidative stress has been associated with a multitude of adverse outcomes in infants, including ROP, BPD, and brain injury (including intraventricular hemorrhage and periventricular leukomalacia). These concerns have led to improvements in methods for measuring oxygenation in neonates, enabling a more precise titration of oxygen delivery. However, the optimal oxygen saturation target in extremely premature infants remains uncertain given the varying results in both randomized and observational studies [66]. Trials conducted in the 1950s demonstrated that administration of high FiO_2 to preterm infants significantly increased their risk of severe ROP and blindness and led to the practice of restricting FiO_2 in the 1960s to no more than 50% [45,67–70]. This change was estimated to result in an excess of 16 deaths per case of blindness prevented [46]. The introduction of transcutaneous PO_2 electrodes allowed for more precise and tighter control of oxygen delivery to preterm infants, and a reduction in ROP was seen in all but the lowest gestational ages. In the 1980s and 1990s, pulse oximetry became the preferred method for monitoring oxygenation, while in the early 2000s, the American Academy of Pediatrics (AAP) suggested a target SpO_2 range of 85–95%, which corresponds to PaO_2 range of 29–67mm Hg (3.8–8.9 kPa), in oxygen-dependent preterm babies in the first 2 weeks of life [71].

3.2. Targeted Use of Supplemental Oxygen

In 2003, the Neonatal Oxygenation Prospective Meta-Analysis (NeOProM) collaborative study was formed to address the question of ideal oxygen targets in extremely preterm infants. The goal was to examine the effects of low versus high functional oxygen saturation targets in the postnatal period in premature infants <28 weeks gestation. Investigators in separate randomized clinical trials prospectively planned to undertake individual trials using similar study designs, participants, interventions, comparators, and outcomes. They agreed to provide individual participant data at trial completion for inclusion in a meta-analysis [69]. Prior to this collaborative effort, there existed only a few small randomized trials in the 1950s [44,72–74] and in the early 2000s [75] investigating oxygen targets in preterm infants. NeOProM ultimately included five multicenter studies under five study groups: Surfactant, Positive Pressure, and Pulse Oximetry Randomized Trial (SUPPORT) in the United States [76]; Benefits of Oxygen Saturation Targeting (BOOST II) trials conducted separately in the United Kingdom, Australia, and New Zealand [77–79]; and the Canadian Oxygen Trial (COT) [80]. In these 5 trials, a total of 4965 premature infants (<28 weeks gestation) were randomized to either a low (85–89%) or high (91–95%) SpO_2 within the first 24 h after birth. Subsequent meta-analyses including follow-up data from the NEOPROM studies have since been published [69,81–83].

The primary outcome for each of the trials was a composite of death or disability by 18–24 months corrected age. No differences in the primary outcome were found between infants assigned to the low (85–89%) or high (91–95%) SpO_2 range. Sub-group analysis also showed no differences for the primary outcome (death or major disability) for gestational age, inborn vs. outborn status, antenatal corticosteroid use, sex, small for gestational age (SGA) vs. appropriate for gestational age (AGA), singleton vs. multiple birth, type of delivery, age at the start of the intervention, or oximeter software type [84].

Despite finding no difference in primary outcome, secondary outcomes from the NeOProM trial raised concerns regarding the importance of oxygen exposure. A higher rate of mortality before the corrected age of 36 weeks, before hospital discharge, and prior to reaching a corrected age of 18 to 24 months was found in infants assigned to the low SpO_2 (85–89%) target range [69,85]. A higher incidence of necrotizing enterocolitis (NEC) was also associated with the low SpO_2 (85–89%) target range. In contrast, there was a lower incidence of ROP requiring treatment associated with the low SpO_2 (85–89%) target range [69,81–84]. Of note, this finding showed substantial heterogeneity because SUPPORT was the only trial within NeOProM to find significant reductions in the need for ROP treatment [85]. Additionally, the increased incidence of ROP in the higher oxygen group (91–95%) did not translate to increased severe visual impairment (defined as bilateral legal blindness) at 18 to 24 months [83]. Additional secondary outcomes including physiologic BPD, intraventricular hemorrhage (IVH), periventricular leukomalacia (PVL), and neurodevelopmental outcomes as assessed by using the Gross Motor Function Classification System or Bayley Scales of Infant Development (BSID) did not differ between the low- and high-oxygen groups [81,82,86].

One main limitation of the NeoProM trial was that significant overlap was found in the actual exposed oxygen saturation between the two groups, despite the design of the trials specifying separation between the target ranges. Although the protocols specified a distinct separation of the SpO_2 target in the two groups there was significant overlap in the SpO_2 achieved, resulting in poor separation between the intervention and comparison groups [82]. Poststudy analysis showed that subjects randomized to 91–95% spent 13.9–22.4% of the time with SpO_2 >95% and subjects randomized to 85–89% spent 20.2–27.4% of the time <85% while on supplemental oxygen [83]. An additional limitation to the NeOProM trial was that two trials (BOOST II trials in the United Kingdom and Australia) were stopped early, which may have resulted in some overestimation of the effect on mortality in these trials [84].

Following the NeOProM publications, some experts in the field suggested that functional SpO_2 should be targeted at 90–95% in infants with gestational age <28 weeks until 36 weeks corrected age and alarm limits 89–95% [71,81]. In 2016, the AAP released new guidelines stating that "the ideal physiologic target range for oxygen saturation for infants of extremely low birth weight is likely patient-specific, dynamic, and dependent on various factors, including gestational age, chronological age, underlying disease, and transfusion status" [86]. Additionally, these guidelines conclude that a target oxygen saturation "range of 90% to 95% may be safer than 85% to 89%, at least for some infants." The 2019 European guidelines recommend "in preterm babies receiving oxygen, the saturation target should be between 90 and 94%" and that "alarm limits should be set to 89 and 95%" [87].

Both the SUPPORT and BOOST trials used fixed, binary SpO_2 targets for the duration of study enrollment (birth to 36 weeks corrected age). It is possible that optimal oxygen saturation varies during the course of NICU hospitalization. The pathophysiology of ROP is thought to involve initial hyperoxic arrest of vascular growth followed by hypoxia-induced hyperproliferation of the retinal vasculature [13,88,89]. In the STOP-ROP trial, infants were enrolled at the time of diagnosis of ROP (mean 10 weeks of absolute age or 35 weeks corrected age) to receive oxygen to target SpO_2 to 89–94% or 96–99%. While the higher oxygen saturation target was associated with adverse pulmonary events and prolonged hospital stay, it was not associated with progression of ROP [75]. In a retrospective study comparing a static SpO_2 goal to a biphasic goal that increased the target to >95% at 34 weeks corrected, infants in the biphasic group had a lower incidence and severity of ROP without an impact on mortality [90].

3.3. Persistent Pulmonary Hypertension of the Newborn

One notable exception to the targeting of specific SpO_2 in the newborn is for infants with persistent pulmonary hypertension of the newborn (PPHN). Prenatally, most of the fetal blood flow bypasses the lungs through elevated pulmonary blood pressure and right-to-left shunts at the level of the atria through the foramen ovale and the great arteries through the ductus arteriosus. During normal birth, crying and initiation of the first breaths results in a rapid decrease in pulmonary blood pressure, allowing blood to circulate through the lungs and participate in gas exchange to oxygenate the infant. When infants do not undergo this physiologic decrease in pulmonary blood pressure, they develop PPHN.

The severity of PPHN can range from mild, requiring only brief oxygen supplementation, to severe, requiring intubation, mechanical ventilation, and even cardiopulmonary bypass with extracorporeal membranous oxygenation (ECMO) [91]. Oxygen is a potent pulmonary vasodilator within the context of hypoxia [92]; as such, up to 100% FiO_2 is commonly used in the treatment of PPHN, regardless of infant SpO_2 [93]. However, there is no clear benefit and potential harm in using such high FiO_2 in the treatment of PPHN. In a lamb model of meconium-aspiration-induced PPHN, pulmonary blood flow and brain oxygen delivery were maximized when targeting a SpO_2 of 95–99% during a 6 h exposure when compared to a fixed FiO_2 of 100% or targeting SpO_2 of 85–89% or 90–94%. Notably, the 95–99% target group showed a mean FiO_2 of 50% and significantly higher pulmonary blood flow than the 100% FiO_2 group [94]. In addition, it is likely that the vasodilatory effect of hyperoxia is short-lived. In fetal lambs exposed to 100% FiO_2, hyperoxic pulmonary vasodilation peaked at approximately 50 min and decreased therafter [95]. Lastly, alveolar hyperoxia may increase pulmonary vascular contractility and impair responsiveness to pulmonary vasodilators, most notably inhaled nitric oxide, worsening outcomes in PPHN [96,97]. The optimal SpO_2 in infants with PPHN remains unclear and requires further study.

4. Future Directions

4.1. Limits of Pulse Oximetry

All of the above studies rely on peripheral pulse oximetry measurements to guide supplemental oxygen administration. Refinements to oximeters and their software have improved portability and reliability; however, they continue to provide only a single data point (arterial oxygen saturation) to drive clinical decision making. The above studies were designed to optimize tissue oxygen delivery without exposing infants to excess oxygen; however, optimal oxgyen delivery depends on other factors, including hemoglobin concentration, tissue oxygen consumption, and hemodynamic status, among a multitude of others [98,99].

In addition, preterm infants have predominantly fetal hemoglobin (HbF), which binds more avidly to oxygen than the predominant adult form of hemoglobin (HbA) [100]. The difference in relative oxygen affinity affects oxyhemoglobin dissociation and delivery to the tissues; thus, an infant with 95% oxyhemoglobin and predominantly HbF may deliver less oxygen to the tissues than an infant with 90% oxyhemoglobin and predominantly HbA. While the ratio of HbF-to-HbA is highly associated with gestational age, interventions to improve oxygen delivery, such as blood transfusions (which are by necessity adult blood), can have a large impact on the amounts of each [101].

4.2. Near-Infrared Spetroscopy

One technology that provides data for other aspects of tissue oxygen delivery is near-infrared spectroscopy (NIRS). As with pulse oximetry, NIRS calculates the ratio of oxyhemoglobin to deoxyhemoglobin, but it does not identify the pulse waveform; therefore, it includes non-arterial sources of transmitted light from the monitored region [102,103]. Because they are measurements of regional oxygen saturation, NIRS measurements are affected by both oxygen consumption and oxygen delivery. The relative contributions of arterial (oxygen delivery) and venous (oxygen consumption) oxygen saturations are

30% and 70%, respectively [104,105]. Cerebral regional oxygen saturation (CrSO$_2$) values have been found to be closely correlated with central venous oxygen saturation, a measure commonly used to calculate global oxygen consumption [106]. There are no established standards for use in neonates; however, multiple prior and ongoing research projects have investigated its use in monitoring regional oxygen saturation, most commonly in the brain and kidneys, and have proposed possible uses for these data [106–110].

A pilot study of very preterm (<32 week) infants using the combination of peripheral pulse oximetry and CrSO$_2$ found significant differences in combined values between infants who developed IVH between 12 and 72 h of life and those who did not [111]. Another study looking specifically at the newborn transition period (first 15 min of life) in very preterm infants found significantly lower CrSO$_2$ in infants who went on to develop IVH in the first 14 days of life [112]. A European unblinded, randomized, controlled trial using CrSO$_2$ to guide therapy, entitled Safeguarding the Brain of Our Smallest Children (SafeBoosC), set a target range for CrSO$_2$ of 55% to 85% and provided 8 treatment options when CrSO$_2$ went below 55%. Notably, only one option involved increasing FiO$_2$ [113]. The trial found a decrease in cerebral hypoxia (CrSO$_2$ < 55%) but no differences in rates of brain injury or neurodevelopmental outcomes at 2 years of age [114]. A systematic review of CrSO$_2$-guided therapy trials found only one trial that met the quality parameters set by reviewers and did not show a reduction in brain injury with the use of cerebral NIRS monitoring [115].

Renal regional oxygen saturation (RrSO$_2$) may also provide enhanced data to guide therapy. It is possible that RrSO$_2$ changes more rapidly than CrSO$_2$ in response to discrepant oxygen delivery and oxygen consumption, as cerebral autoregulation protects the brain against rapid changes in perfusion in response to changes in hemodynamics [116]. Interestingly, higher RrSO$_2$ in neonates with hypoxic ischemic encephalopathy was shown to be associated with acute kidney injury [117]. In contrast, in infants undergoing heart surgery, low RrSO$_2$ was associated with the development of acute kidney injury [118].

Specific guidelines for the use and interpretation of NIRS may provide a framework to apply this technology to specific clinical scenarios to optimize oxygen delivery in the NICU [119]. The simplicity of pulse oximetry makes it an ideal monitor to guide clinical therapy, and the refinement of oxygen therapy guidelines using pulse oximetry has clearly shown benefit. The question remains whether the precision afforded by the addition of NIRS monitoring has enough benefit to overcome the burden of additional monitors on fragile infants in the NICU with limited body surface area.

4.3. Novel Therapies

In addition to limitations in monitoring, there are no available therapies to mitigate the effects of oxidative injury in the NICU. In a review of antioxidant therapies with particular focus on BPD, Ofman and Tipple outlined major barriers to effective therapies, including lack of compartment and target specificity, limited bioavailability, timing of therapy, and genetic variability [120]. Endogenous antioxidant enzymes are highly localized and perform specific functions in redox biochemistry. Systemic or even organ-specific administration of antioxidants may not reach the cells or organelles where hyperoxia induces injury. In addition, individual organs have variable oxygen consumption rates and respond differently to hypoxia and hyperoxia. This is evident in the NeOProM trials, where lower SpO$_2$ targets were associated with NEC and mortality, while higher SpO$_2$ targets were associated with ROP. Future therapies could target oxygen delivery specifically to the organs most in need. For example, a bio-engineered heme-containing protein, OMX-CV, has a 10-fold higher affinity for oxygen than hemoglobin does and has been shown to selectively deliver oxygen to hypoxic tissues but not to tissues at physiologic oxygen tension rates in juvenile lambs [121].

Investigations of the redox signaling pathways involved in oxidative stress have revealed other potential future targets for clinical study. Thioredoxin is an antioxidant enzyme that catalyzes the reduction of oxidized cysteine residues, playing an important role in reducing protein disulfide bonds. Alterations in thioredoxin and its partner enzyme, thioredoxin reductase, have been shown to affect susceptibility to hyperoxic lung injury [122]. Interestingly, the drugs auranofin and aurothioglucose inhibit thioredoxin reductase and have been shown to protect mice from hyperoxic lung injury through the induction of nuclear factor erythroid 2-related factor (NRF2)-induced genes [123,124]. NRF2 is an oxidant-activated transcription factor that stimulates the transcription of a variety of endogenous antioxidant systems [125]. Similarly, melatonin is a neurohormone that induces a variety of endogenous antioxidant enzymes and has been shown to mitigate hyperoxic lung injury in rats, as well as hypoxic ischemic encephalopathy in pigs [126–128]. In contrast to trials involving the direct application of antioxidant enzymes or cofactors, these medications provide an alternative approach through the induction of endogenous antioxidant systems.

Lastly, patient responses to hypoxia and hyperoxia may vary based on genetic differences between antioxidant systems. In a population of very low birth weight infants, sequencing of a variety of antioxidant genes revealed single-nucleotide polymorphisms in the NQO1 gene (coding for NADPH quinone reductase) and the NFE2L2 gene (coding for NRF2), which were associated significant differences in the risk of BPD [129]. As exome and genome sequencing become cheaper and more ubiquitous in the NICU, identification of genetic differences in endogenous antioxidant genes may allow for individualized approaches to mitigating oxidative injury.

5. Conclusions

Preterm and term newborns are unique populations with decreased antioxidant capacity and dependence on redox signaling for rapid growth and development. Despite advances in neonatal care, supplemental oxygen remains the most commonly used drug in the NICU. While oxygen can be life-saving, it is toxic in excess. Its application to neonatal medicine has seen a dramatic pendulum swing, from ubiquitous use both in the delivery room and beyond to intense restriction due to concern for retinal disease. The introduction of the pulse oximeter to initial newborn resuscitation and the NICU allowed for targeted oxygen therapy, while broad international collaboration has resulted in ample data to support current guidelines. Nonetheless, pulse oximetry provides limited information with which to make therapeutic decisions, and supplemental oxygen remains a blunt and untargeted therapy. Future strategies to optimize oxygen delivery while mitigating oxidative damage may involve NIRS monitoring or organ-, cell-, and patient-specific therapies that harness endogenous antioxidant systems.

Author Contributions: M.M., J.C., M.P. and O.S. conceptualized the content and wrote, edited, and approved the manuscript. All authors have read and agreed to the published version of the manuscript.

Funding: This research was funded by the National Institutes of Health to J.C. (K08 HD102023) and M.P. (K08 HL124295).

Institutional Review Board Statement: Not applicable.

Informed Consent Statement: Not applicable.

Data Availability Statement: Not applicable.

Conflicts of Interest: The authors declare no conflict of interest.

References

1. Johnson, N.; Johnson, V.A.; McNamara, H.; Montague, I.A.; Jongsma, H.W.; Aumeerally, Z.; Gupta, J.K.; van Oudgaarden, E.; Lilford, R.J.; Miller, D. Fetal Pulse Oximetry: A New Method of Monitoring the Fetus. *Aust. N. Z. J. Obstet. Gynaecol.* **1994**, *34*, 428–432. [CrossRef]
2. Butte, N.F.; Moon, J.K.; Wong, W.W.; Hopkinson, J.M.; Smith, E.O. Energy Requirements from Infancy to Adulthood. *Am. J. Clin. Nutr.* **1995**, *62*, 1047S–1052S. [CrossRef]

3. Vannucci, S.J. Developmental Expression of GLUT1 and GLUT3 Glucose Transporters in Rat Brain. *J. Neurochem.* **1994**, *62*, 240–246. [CrossRef]
4. Ten, V.S.; Stepanova, A.A.; Ratner, V.; Neginskaya, M.; Niatsetskaya, Z.; Sosunov, S.; Starkov, A. Mitochondrial Dysfunction and Permeability Transition in Neonatal Brain and Lung Injuries. *Cells* **2021**, *10*, 569. [CrossRef]
5. Finkel, T. Signal Transduction by Mitochondrial Oxidants. *J. Biol. Chem.* **2012**, *287*, 4434–4440. [CrossRef] [PubMed]
6. Berkelhamer, S.K.; Kim, G.A.; Radder, J.E.; Wedgwood, S.; Czech, L.; Steinhorn, R.H.; Schumacker, P.T. Developmental Differences in Hyperoxia-Induced Oxidative Stress and Cellular Responses in the Murine Lung. *Free Radic. Biol. Med.* **2013**, *61*, 51–60. [CrossRef] [PubMed]
7. Saugstad, O.D.; Hallman, M.; Abraham, J.L.; Epstein, B.; Cochrane, C.; Gluck, L. Hypoxanthine and Oxygen Induced Lung Injury: A Possible Basic Mechanism of Tissue Damage? *Pediatr. Res.* **1984**, *18*, 501–504. [CrossRef]
8. Ogihara, T.; Hirano, K.; Morinobu, T.; Kim, H.-S.; Hiroi, M.; Ogihara, H.; Tamai, H. Raised Concentrations of Aldehyde Lipid Peroxidation Products in Premature Infants with Chronic Lung Disease. *Arch. Dis. Child. Fetal Neonatal. Ed.* **1999**, *80*, F21–F25. [CrossRef]
9. Warner, B.B.; Stuart, L.A.; Papes, R.A.; Wispé, J.R. Functional and Pathological Effects of Prolonged Hyperoxia in Neonatal Mice. *Am. J. Physiol. -Lung Cell. Mol. Physiol.* **1998**, *275*, L110–L117. [CrossRef] [PubMed]
10. Ratner, V.; Starkov, A.; Matsiukevich, D.; Polin, R.A.; Ten, V.S. Mitochondrial Dysfunction Contributes to Alveolar Developmental Arrest in Hyperoxia-Exposed Mice. *Am. J. Respir. Cell Mol. Biol.* **2009**, *40*, 511–518. [CrossRef]
11. Giannone, P.; Bauer, J.; Schanbacher, B.; Reber, K. Effects of Hyperoxia on Postnatal Intestinal Development. *Biotech. Histochem.* **2007**, *82*, 17–22. [CrossRef]
12. Liu, D.Y.; Lou, W.J.; Zhang, D.Y.; Sun, S.Y. ROS Plays a Role in the Neonatal Rat Intestinal Barrier Damages Induced by Hyperoxia. *BioMed Res. Int.* **2020**, *2020*, e8819195. [CrossRef]
13. Hartnett, M.E.; Penn, J.S. Mechanisms and Management of Retinopathy of Prematurity. *N. Engl. J. Med.* **2012**, *367*, 2515–2526. [CrossRef]
14. Kennedy, C.; Grave, G.D.; Sokoloff, L. Alterations of Local Cerebral Blood Flow Due to Exposure of Newborn Puppies to 80–90 per Cent Oxygen. *Eur. Neurol.* **1971**, *6*, 137–140. [CrossRef] [PubMed]
15. Liu, Y.; Jiang, P.; Du, M.; Chen, K.; Chen, A.; Wang, Y.; Cao, F.; Deng, S.; Xu, Y. Hyperoxia-Induced Immature Brain Injury through the TLR4 Signaling Pathway in Newborn Mice. *Brain Res.* **2015**, *1610*, 51–60. [CrossRef]
16. Du, M.; Tan, Y.; Liu, G.; Liu, L.; Cao, F.; Liu, J.; Jiang, P.; Xu, Y. Effects of the Notch Signalling Pathway on Hyperoxia-Induced Immature Brain Damage in Newborn Mice. *Neurosci. Lett.* **2017**, *653*, 220–227. [CrossRef]
17. Lee, Y.-S.; Chou, Y.-H. Antioxidant Profiles in Full Term and Preterm Neonates. *Chang. Gung Med. J.* **2005**, *28*, 846–851. [PubMed]
18. Heinonen, K.; Mononen, I.; Mononen, T.; Parviainen, M.; Penttilä, I.; Launiala, K. Plasma Vitamin C Levels Are Low in Premature Infants Fed Human Milk. *Am. J. Clin. Nutr* **1986**, *43*, 923–924. [CrossRef] [PubMed]
19. Fort, P.; Salas, A.A.; Nicola, T.; Craig, C.M.; Carlo, W.A.; Ambalavanan, N. A Comparison of 3 Vitamin D Dosing Regimens in Extremely Preterm Infants: A Randomized Controlled Trial. *J. Pediatr.* **2016**, *174*, 132–138.e1. [CrossRef]
20. Falciglia, H.S.; Ginn-Pease, M.E.; Falciglia, G.A.; Lubin, A.H.; Frank, D.J.; Chang, W. Vitamin E and Selenium Levels of Premature Infants with Severe Respiratory Distress Syndrome and Bronchopulmonary Dysplasia. *J. Pediatr. Perinat Nutr.* **1988**, *2*, 35–49. [CrossRef]
21. Rosenfeld, W.; Evans, H.; Concepcion, L.; Jhaveri, R.; Schaeffer, H.; Friedman, A. Prevention of Bronchopulmonary Dysplasia by Administration of Bovine Superoxide Dismutase in Preterm Infants with Respiratory Distress Syndrome. *J. Pediatr.* **1984**, *105*, 781–785. [CrossRef]
22. Davis, J.M.; Rosenfeld, W.N.; Richter, S.E.; Parad, R.; Gewolb, I.H.; Spitzer, A.R.; Carlo, W.A.; Couser, R.J.; Price, A.; Flaster, E.; et al. Safety and Pharmacokinetics of Multiple Doses of Recombinant Human CuZn Superoxide Dismutase Administered Intratracheally to Premature Neonates with Respiratory Distress Syndrome. *Pediatrics* **1997**, *100*, 24–30. [CrossRef]
23. Davis, J.M.; Parad, R.B.; Michele, T.; Allred, E.; Price, A.; Rosenfeld, W. Pulmonary Outcome at 1 Year Corrected Age in Premature Infants Treated at Birth with Recombinant Human CuZn Superoxide Dismutase. *Pediatrics* **2003**, *111*, 469–476. [CrossRef]
24. Suresh, G.; Davis, J.M.; Soll, R. Superoxide Dismutase for Preventing Chronic Lung Disease in Mechanically Ventilated Preterm Infants. *Cochrane Database Syst. Rev.* **2001**, *2001*, CD001968. [CrossRef] [PubMed]
25. Meyer, S.; Gortner, L.; Investigators, N.T. Early Postnatal Additional High-Dose Oral Vitamin A Supplementation versus Placebo for 28 Days for Preventing Bronchopulmonary Dysplasia or Death in Extremely Low Birth Weight Infants. *Neonatology* **2014**, *105*, 182–188. [CrossRef] [PubMed]
26. Meyer, S.; Gortner, L.; Meyer, S.; Bay, J.; Gortner, L.; Ehrlich, A.; Ruckes, C.; Seidenberg, J.; Muyimbwa, J.; Wieg, C.; et al. Up-Date on the NeoVitaA Trial: Obstacles, Challenges, Perspectives, and Local Experiences. *Wien. Med. Wochenschr.* **2017**, *167*, 264–270. [CrossRef] [PubMed]
27. Tyson, J.E.; Kennedy, K.A.; Stoll, B.J.; Bauer, C.R. Vitamin A Supplementation for Extremely-Low-Birth-Weight Infants. *N. Engl. J. Med.* **1999**, *7*. [CrossRef]
28. Darlow, B.A.; Graham, P.J.; Rojas-Reyes, M.X. Vitamin A Supplementation to Prevent Mortality and Short- and Long-term Morbidity in Very Low Birth Weight Infants. *Cochrane Database Syst. Rev.* **2016**, *2016*, CD000501. [CrossRef]

29. Kiatchoosakun, P.; Jirapradittha, J.; Panthongviriyakul, M.C.; Khampitak, T.; Yongvanit, P.; Boonsiri, P. Vitamin A Supplementation for Prevention of Bronchopulmonary Dysplasia in Very-Low-Birth-Weight Premature Thai Infants: A Randomized Trial. *J. Med. Assoc. Thai* **2014**, *97* (Suppl. S10), S82–S88.

30. Darlow, B.A.; Buss, H.; McGill, F.; Fletcher, L.; Graham, P.; Winterbourn, C.C. Vitamin C Supplementation in Very Preterm Infants: A Randomised Controlled Trial. *Arch. Dis Child. Fetal Neonatal Ed.* **2005**, *90*, F117–F122. [CrossRef]

31. Brion, L.P.; Bell, E.F.; Raghuveer, T.S. Vitamin E Supplementation for Prevention of Morbidity and Mortality in Preterm Infants. *Cochrane Database Syst. Rev.* **2003**, *2003*, CD003665. [CrossRef]

32. Stone, C.A., Jr.; McEvoy, C.T.; Aschner, J.L.; Kirk, A.; Rosas-Salazar, C.; Cook-Mills, J.M.; Moore, P.E.; Walsh, W.F.; Hartert, T.V. Update on Vitamin E and Its Potential Role in Preventing or Treating Bronchopulmonary Dysplasia. *Neonatology* **2018**, *113*, 366–378. [CrossRef] [PubMed]

33. Watts, J.L.; Milner, R.; Zipursky, A.; Paes, B.; Ling, E.; Gill, G.; Fletcher, B.; Rand, C. Failure of Supplementation with Vitamin E to Prevent Bronchopulmonary Dysplasia in Infants Less than 1500 g Birth Weight. *Eur. Respir. J.* **1991**, *4*, 188–190. [PubMed]

34. Ahola, T.; Lapatto, R.; Raivio, K.O.; Selander, B.; Stigson, L.; Jonsson, B.; Jonsbo, F.; Esberg, G.; Stövring, S.; Kjartansson, S.; et al. N-Acetylcysteine Does Not Prevent Bronchopulmonary Dysplasia in Immature Infants: A Randomized Controlled Trial. *J. Pediatr.* **2003**, *143*, 713–719. [CrossRef]

35. Rhee, S.G.; Bae, Y.S.; Lee, S.-R.; Kwon, J. Hydrogen Peroxide: A Key Messenger That Modulates Protein Phosphorylation Through Cysteine Oxidation. *Sci. Signal.* **2000**, *2000*, pe1. [CrossRef] [PubMed]

36. Salmeen, A.; Andersen, J.N.; Myers, M.P.; Meng, T.-C.; Hinks, J.A.; Tonks, N.K.; Barford, D. Redox Regulation of Protein Tyrosine Phosphatase 1B Involves a Sulphenyl-Amide Intermediate. *Nature* **2003**, *423*, 769–773. [CrossRef]

37. Lee, S.-R.; Kwon, K.-S.; Kim, S.-R.; Rhee, S.G. Reversible Inactivation of Protein-Tyrosine Phosphatase 1B in A431 Cells Stimulated with Epidermal Growth Factor*. *J. Biol. Chem.* **1998**, *273*, 15366–15372. [CrossRef]

38. Oshikawa, J.; Urao, N.; Kim, H.W.; Kaplan, N.; Razvi, M.; McKinney, R.; Poole, L.B.; Fukai, T.; Ushio-Fukai, M. Extracellular SOD-Derived H_2O_2 Promotes VEGF Signaling in Caveolae/Lipid Rafts and Post-Ischemic Angiogenesis in Mice. *PLoS ONE* **2010**, *5*, e10189. [CrossRef]

39. Nezu, M.; Souma, T.; Yu, L.; Sekine, H.; Takahashi, N.; Wei, A.Z.-S.; Ito, S.; Fukamizu, A.; Zsengeller, Z.K.; Nakamura, T.; et al. Nrf2 Inactivation Enhances Placental Angiogenesis in a Preeclampsia Mouse Model and Improves Maternal and Fetal Outcomes. *Sci. Signal.* **2017**, *10*, eaam5711. [CrossRef]

40. Flagg, P.J. Treatment of Asphyxia in the New-Born: Preliminary Report of the Practical Application of Modern Scientific Methods. *J. Am. Med. Assoc.* **1928**, *91*, 788–791. [CrossRef]

41. Terry, T.L. Extreme Prematurity and Fibroblastic Overgrowth of Persistent Vascular Sheath Behind Each Crystalline Lens: I. Preliminary Report. *Am. J. Ophthalmol.* **2018**, *192*, xxviii. [CrossRef]

42. Owens, W.C.; Owens, E.U. Retrolental Fibroplasia in Premature Infants. *Trans. Am. Acad. Ophthalmol. Otolaryngol.* **1948**, *53*, 18–41.

43. Appelbaum, A. Retrolental Fibroplasia—Blindness in Infants of Low Weight at Birth. *Calif. Med.* **1952**, *77*, 259–265.

44. Kinsey, V.E. Retrolental Fibroplasia; Cooperative Study of Retrolental Fibroplasia and the Use of Oxygen. *AMA Arch. Ophthalmol.* **1956**, *56*, 481–543. [CrossRef]

45. Campbell, K. Intensive Oxygen Therapy as a Possible Cause of Retrolental Fibroplasia; a Clinical Approach. *Med. J. Aust.* **1951**, *2*, 48–50. [CrossRef]

46. Cross, K.W. Cost of preventing retrolental fibroplasia? *Lancet* **1973**, *302*, 954–956. [CrossRef]

47. Lucey, J.F.; Dangman, B. A Reexamination of the Role of Oxygen in Retrolental Fibroplasia. *Pediatrics* **1984**, *73*, 82–96.

48. Klaus, M.; Meyer, B.P. Oxygen Therapy for the Newborn. *Pediatr. Clin. N. Am.* **1966**, *13*, 731–752. [CrossRef]

49. Kerber, R.; Ornato, J.; Brown, D.; Chameides, L.; Chandra, N.; Cummins, R.; Hazinski, M.; Melker, R.; Weaver, D. Emergency Cardiac Care Committee and Subcommittees, American Heart Association. *JAMA* **1992**, *268*, 2171. [CrossRef]

50. Saugstad, O.D.; Gluck, L. Plasma Hypoxanthine Levels in Newborn Infants: A Specific Indicator of Hypoxia. *J. Perinat Med.* **1982**, *10*, 266–272. [CrossRef] [PubMed]

51. Feet, B.A.; Yu, X.-Q.; Rootwelt, T.; Oyasaeter, S.; Saugstad, O.D. Effects of Hypoxemia and Reoxygenation with 21% or 100% Oxygen in Newborn Piglets: Extracellular Hypoxanthine in Cerebral Cortex and Femoral Muscle. *Crit. Care Med.* **1997**, *25*, 1384–1391. [CrossRef]

52. Rootwelt, T.; Løberg, E.M.; Moen, A.; Øyasæter, S.; Saugstad, O.D. Hypoxemia and Reoxygenation with 21% or 100% Oxygen in Newborn Pigs: Changes in Blood Pressure, Base Deficit, and Hypoxanthine and Brain Morphology. *Pediatr. Res.* **1992**, *32*, 107–113. [CrossRef] [PubMed]

53. Saugstad, O.D.; Rootwelt, T.; Aalen, O. Resuscitation of Asphyxiated Newborn Infants with Room Air or Oxygen: An International Controlled Trial: The Resair 2 Study. *Pediatrics* **1998**, *102*, e1. [CrossRef] [PubMed]

54. Saugstad, O.D.; Ramji, S.; Irani, S.F.; El-Meneza, S.; Hernandez, E.A.; Vento, M.; Talvik, T.; Solberg, R.; Rootwelt, T.; Aalen, O.O. Resuscitation of Newborn Infants with 21% or 100% Oxygen: Follow-Up at 18 to 24 Months. *Pediatrics* **2003**, *112*, 296–300. [CrossRef]

55. Vento, M.; Asensi, M.; Sastre, J.; Garcíia-Sala, F.; Pallardó, F.V.; Viña, J. Resuscitation with Room Air Instead of 100% Oxygen Prevents Oxidative Stress in Moderately Asphyxiated Term Neonates. *Pediatrics* **2001**, *107*, 642–647. [CrossRef] [PubMed]

56. Vento, M.; Asensi, M.; Sastre, J.; García-Sala, F.; Viña, J. Six Years of Experience with the Use of Room Air for the Resuscitation of Asphyxiated Newly Born Term Infants. *Biol. Neonate* **2001**, *79*, 261–267. [CrossRef]

57. Welsford, M.; Nishiyama, C.; Shortt, C.; Isayama, T.; Dawson, J.A.; Weiner, G.; Roehr, C.C.; Wyckoff, M.H.; Rabi, Y. Room Air for Initiating Term Newborn Resuscitation: A Systematic Review with Meta-Analysis. *Pediatrics* **2019**, *143*, e20181825. [CrossRef]
58. Escobedo, M.B.; Aziz, K.; Kapadia, V.S.; Lee, H.C.; Niermeyer, S.; Schmölzer, G.M.; Szyld, E.; Weiner, G.M.; Wyckoff, M.H.; Yamada, N.K.; et al. 2019 American Heart Association Focused Update on Neonatal Resuscitation: An Update to the American Heart Association Guidelines for Cardiopulmonary Resuscitation and Emergency Cardiovascular Care. *Pediatrics* **2020**, *145*, e922–e930. [CrossRef]
59. Wyckoff, M.H.; Wyllie, J.; Aziz, K.; de Almeida, M.F.; Fabres, J.; Fawke, J.; Guinsburg, R.; Hosono, S.; Isayama, T.; Kapadia, V.S.; et al. Neonatal Life Support: 2020 International Consensus on Cardiopulmonary Resuscitation and Emergency Cardiovascular Care Science with Treatment Recommendations. *Circulation* **2020**, *142*, S185–S221. [CrossRef]
60. Lui, K.; Jones, L.J.; Foster, J.P.; Davis, P.G.; Ching, S.K.; Oei, J.L.; Osborn, D.A. Lower versus Higher Oxygen Concentrations Titrated to Target Oxygen Saturations during Resuscitation of Preterm Infants at Birth. *Cochrane Database Syst. Rev.* **2018**, *2018*, CD010239. [CrossRef]
61. Oei, J.L.; Saugstad, O.D.; Lui, K.; Wright, I.M.; Smyth, J.P.; Craven, P.; Wang, Y.A.; McMullan, R.; Coates, E.; Ward, M.; et al. Targeted Oxygen in the Resuscitation of Preterm Infants, a Randomized Clinical Trial. *Pediatrics* **2017**, *139*. [CrossRef] [PubMed]
62. Rabi, Y.; Lodha, A.; Soraisham, A.; Singhal, N.; Barrington, K.; Shah, P.S. Outcomes of Preterm Infants Following the Introduction of Room Air Resuscitation. *Resuscitation* **2015**, *96*, 252–259. [CrossRef] [PubMed]
63. Soraisham, A.S.; Rabi, Y.; Shah, P.S.; Singhal, N.; Synnes, A.; Yang, J.; Lee, S.K.; Lodha, A.K. Neurodevelopmental Outcomes of Preterm Infants Resuscitated with Different Oxygen Concentration at Birth. *J. Perinatol.* **2017**, *37*, 1141–1147. [CrossRef] [PubMed]
64. Thamrin, V.; Saugstad, O.D.; Tarnow-Mordi, W.; Wang, Y.A.; Lui, K.; Wright, I.M.; De Waal, K.; Travadi, J.; Smyth, J.P.; Craven, P.; et al. Preterm Infant Outcomes after Randomization to Initial Resuscitation with FiO$_2$ 0.21 or 1.0. *J. Pediatr.* **2018**, *201*, 55–61.e1. [CrossRef]
65. Kapadia, V.; Oei, J.L.; Finer, N.; Rich, W.; Rabi, Y.; Wright, I.M.; Rook, D.; Vermeulen, M.J.; Tarnow-Mordi, W.O.; Smyth, J.P.; et al. Outcomes of Delivery Room Resuscitation of Bradycardic Preterm Infants: A Retrospective Cohort Study of Randomised Trials of High vs. Low Initial Oxygen Concentration and an Individual Patient Data Analysis. *Resuscitation* **2021**, *167*, 209–217. [CrossRef]
66. Greenspan, J.S.; Goldsmith, J.P. Oxygen Therapy in Preterm Infants: Hitting the Target. *Pediatrics* **2006**, *118*, 1740–1741. [CrossRef]
67. Silverman, W.A. Oxygen Therapy and Retrolental Fibroplasia. *Am. J. Public Health Nations Health* **1968**, *58*, 2009–2011. [CrossRef]
68. Askie, L.M.; Henderson-Smart, D.J.; Ko, H. Restricted versus Liberal Oxygen Exposure for Preventing Morbidity and Mortality in Preterm or Low Birth Weight Infants. *Cochrane Database Syst. Rev.* **2009**, *2009*, CD001077. [CrossRef]
69. Askie, L.M.; Darlow, B.A.; Finer, N.; Schmidt, B.; Stenson, B.; Tarnow-Mordi, W.; Davis, P.G.; Carlo, W.A.; Brocklehurst, P.; Davies, L.C.; et al. Association Between Oxygen Saturation Targeting and Death or Disability in Extremely Preterm Infants in the Neonatal Oxygenation Prospective Meta-Analysis Collaboration. *JAMA* **2018**, *319*, 2190–2201. [CrossRef]
70. Robertson, A.F. Reflections on Errors in Neonatology: I. The "Hands-Off" Years, 1920 to 1950. *J. Perinatol.* **2003**, *23*, 48–55. [CrossRef]
71. Saugstad, O.D. Oxygenation of the Immature Infant: A Commentary and Recommendations for Oxygen Saturation Targets and Alarm Limits. *Neonatology* **2018**, *114*, 69–75. [CrossRef] [PubMed]
72. Lanman, J.T.; Guy, L.P.; Dancis, J. Retrolental Fibroplasia and Oxygen Therapy. *J. Am. Med. Assoc.* **1954**, *155*, 223–226. [CrossRef]
73. Patz, A. Oxygen Studies in Retrolental Fibroplasia*: IV. Clinical and Experimental Observations the First Edward L. Holmes Memorial Lecture. *Am. J. Ophthalmol.* **1954**, *38*, 291–308. [CrossRef]
74. Patz, A. The Role of Oxygen in Retrolental Fibroplasia. *Sinai Hosp. J.* **1954**, *3*, 6–20. [CrossRef] [PubMed]
75. STOP-ROP Multicenter Study Group. Supplemental Therapeutic Oxygen for Prethreshold Retinopathy of Prematurity (STOP-ROP), A Randomized, Controlled Trial. I: Primary Outcomes. *Pediatrics* **2000**, *105*, 295–310. [CrossRef] [PubMed]
76. SUPPORT Study Group of the Eunice Kennedy Shriver NICHD Neonatal Research Network; Carlo, W.A.; Finer, N.N.; Walsh, M.C.; Rich, W.; Gantz, M.G.; Laptook, A.R.; Yoder, B.A.; Faix, R.G.; Das, A.; et al. Target Ranges of Oxygen Saturation in Extremely Preterm Infants. *N. Engl. J. Med.* **2010**, *362*, 1959–1969. [CrossRef] [PubMed]
77. BOOST II United Kingdom Collaborative Group; BOOST II Australia Collaborative Group; BOOST II New Zealand Collaborative Group; Stenson, B.J.; Tarnow-Mordi, W.O.; Darlow, B.A.; Simes, J.; Juszczak, E.; Askie, L.; Battin, M.; et al. Oxygen Saturation and Outcomes in Preterm Infants. *N. Engl. J. Med.* **2013**, *368*, 2094–2104. [CrossRef]
78. Stenson, B.; Brocklehurst, P.; Tarnow-Mordi, W. Increased 36-Week Survival with High Oxygen Saturation Target in Extremely Preterm Infants. *N. Engl. J. Med.* **2011**, *364*, 1680–1682. [CrossRef] [PubMed]
79. BOOST-II Australia and United Kingdom Collaborative Groups; Tarnow-Mordi, W.; Stenson, B.; Kirby, A.; Juszczak, E.; Donoghoe, M.; Deshpande, S.; Morley, C.; King, A.; Doyle, L.W.; et al. Outcomes of Two Trials of Oxygen-Saturation Targets in Preterm Infants. *N. Engl. J. Med.* **2016**, *374*, 749–760. [CrossRef]
80. Al Hazzani, F.; Khadawardi, E. Effects of Targeting Higher vs. Lower Arterial Oxygen Saturations on Death or Disability in Extremely Preterm Infants: The Canadian Oxygen Trial. *J. Clin. Neonatol.* **2013**, *2*, 70–72. [CrossRef]
81. Saugstad, O.D.; Aune, D. Optimal Oxygenation of Extremely Low Birth Weight Infants: A Meta-Analysis and Systematic Review of the Oxygen Saturation Target Studies. *Neonatology* **2014**, *105*, 55–63. [CrossRef] [PubMed]
82. Manja, V.; Lakshminrusimha, S.; Cook, D.J. Oxygen Saturation Target Range for Extremely Preterm Infants: A Systematic Review and Meta-Analysis. *JAMA Pediatr.* **2015**, *169*, 332–340. [CrossRef] [PubMed]

83. Manja, V.; Saugstad, O.D.; Lakshminrusimha, S. Oxygen Saturation Targets in Preterm Infants and Outcomes at 18-24 Months: A Systematic Review. *Pediatrics* **2017**, *139*, e20161609. [CrossRef] [PubMed]

84. Schmidt, B.; Whyte, R.K. Oxygen Saturation Target Ranges and Alarm Settings in the NICU: What Have We Learnt from the Neonatal Oxygenation Prospective Meta-Analysis (NeOProM)? *Semin. Fetal. Neonatal Med.* **2020**, *25*, 101080. [CrossRef] [PubMed]

85. Askie, L.M. Meta-Analysis of Oxygenation Saturation Targeting Trials: Do Infant Subgroups Matter? *Clin. Perinatol.* **2019**, *46*, 579–591. [CrossRef]

86. Kilpatrick, S.J.; American Academy of Pediatrics; American College of Obstetricians and Gynecologists. *Guidelines for Perinatal Care*; American Academy of Pediatrics; The American College of Obstetricians and Gynecologists: Washington, DC, USA, 2017; ISBN 978-1-61002-088-6.

87. Sweet, D.G.; Carnielli, V.; Greisen, G.; Hallman, M.; Ozek, E.; Te Pas, A.; Plavka, R.; Roehr, C.C.; Saugstad, O.D.; Simeoni, U.; et al. European Consensus Guidelines on the Management of Respiratory Distress Syndrome—2019 Update. *Neonatology* **2019**, *115*, 432–450. [CrossRef]

88. Bashinsky, A.L. Retinopathy of Prematurity. *North. Carol. Med. J.* **2017**, *78*, 124–128. [CrossRef]

89. Hartnett, M.E.; Lane, R.H. Effects of Oxygen on the Development and Severity of Retinopathy of Prematurity. *J. Am. Assoc. Pediatr. Ophthalmol. Strabismus* **2013**, *17*, 229–234. [CrossRef]

90. Shukla, A.; Sonnie, C.; Worley, S.; Sharma, A.; Howard, D.; Moore, J.; Rodriguez, R.J.; Hoppe, G.; Sears, J.E. Comparison of Biphasic vs. Static Oxygen Saturation Targets Among Infants with Retinopathy of Prematurity. *JAMA Ophthalmol.* **2019**, *137*, 417–423. [CrossRef]

91. Sharma, M.; Mohan, K.R.; Narayan, S.; Chauhan, L. Persistent Pulmonary Hypertension of the Newborn: A Review. *Med. J. Armed Forces India* **2011**, *67*, 348–353. [CrossRef]

92. Rudolph, A.M.; Yuan, S. Response of the Pulmonary Vasculature to Hypoxia and H+ Ion Concentration Changes. *J. Clin. Investig.* **1966**, *45*, 399–411. [CrossRef] [PubMed]

93. Nair, J.; Lakshminrusimha, S. Update on pphn: Mechanisms and treatment. *Semin. Perinatol.* **2014**, *38*, 78–91. [CrossRef] [PubMed]

94. Rawat, M.; Chandrasekharan, P.; Gugino, S.F.; Koenigsknecht, C.; Nielsen, L.; Wedgwood, S.; Mathew, B.; Nair, J.; Steinhorn, R.; Lakshminrusimha, S. Optimal Oxygen Targets in Term Lambs with Meconium Aspiration Syndrome and Pulmonary Hypertension. *Am. J. Respir Cell Mol. Biol* **2020**, *63*, 510–518. [CrossRef] [PubMed]

95. Accurso, F.J.; Alpert, B.; Wilkening, R.B.; Petersen, R.G.; Meschia, G. Time-Dependent Response of Fetal Pulmonary Blood Flow to an Increase in Fetal Oxygen Tension. *Respir. Physiol.* **1986**, *63*, 43–52. [CrossRef]

96. Lakshminrusimha, S.; Russell, J.A.; Steinhorn, R.H.; Ryan, R.M.; Gugino, S.F.; Morin, F.C.; Swartz, D.D.; Kumar, V.H. Pulmonary Arterial Contractility in Neonatal Lambs Increases with 100% Oxygen Resuscitation. *Pediatr. Res.* **2006**, *59*, 137–141. [CrossRef]

97. Lakshminrusimha, S.; Swartz, D.D.; Gugino, S.F.; Ma, C.-X.; Wynn, K.A.; Ryan, R.M.; Russell, J.A.; Steinhorn, R.H. Oxygen Concentration and Pulmonary Hemodynamics in Newborn Lambs with Pulmonary Hypertension. *Pediatr. Res.* **2009**, *66*, 539–544. [CrossRef]

98. Dunn, J.-O.; Mythen, M.; Grocott, M. Physiology of Oxygen Transport. *BJA Educ.* **2016**, *16*, 341–348. [CrossRef]

99. Whyte, R.K.; Jangaard, K.A.; Dooley, K.C. From Oxygen Content to Pulse Oximetry: Completing the Picture in the Newborn. *Acta Anaesthesiol. Scand. Suppl.* **1995**, *107*, 95–100. [CrossRef]

100. Wilson, K.; Hawken, S.; Murphy, M.S.Q.; Atkinson, K.M.; Potter, B.K.; Sprague, A.; Walker, M.; Chakraborty, P.; Little, J. Postnatal Prediction of Gestational Age Using Newborn Fetal Hemoglobin Levels. *EBioMedicine* **2017**, *15*, 203–209. [CrossRef]

101. De Halleux, V.; Truttmann, A.; Gagnon, C.; Bard, H. The Effect of Blood Transfusion on the Hemoglobin Oxygen Dissociation Curve of Very Early Preterm Infants during the First Week of Life. *Semin. Perinatol.* **2002**, *26*, 411–415. [CrossRef]

102. Whetsel, K.B. Near-Infrared Spectrophotometry. *Appl. Spectrosc. Rev.* **1968**, *2*, 1–67. [CrossRef]

103. Suzuki, S.; Takasaki, S.; Ozaki, T.; Kobayashi, Y. Tissue Oxygenation Monitor Using NIR Spatially Resolved Spectroscopy. In Proceedings of the Optical Tomography and Spectroscopy of Tissue III, San Jose, CA, USA, 24–28 January 1999; Volume 3597, pp. 582–592.

104. Watzman, H.M.; Kurth, C.D.; Montenegro, L.M.; Rome, J.; Steven, J.M.; Nicolson, S.C. Arterial and Venous Contributions to Near-Infrared Cerebral Oximetry. *Anesthesiology* **2000**, *93*, 947–953. [CrossRef] [PubMed]

105. Wong, F.Y.; Alexiou, T.; Samarasinghe, T.; Brodecky, V.; Walker, A.M. Cerebral Arterial and Venous Contributions to Tissue Oxygenation Index Measured Using Spatially Resolved Spectroscopy in Newborn Lambs. *Anesthesiology* **2010**, *113*, 1385–1391. [CrossRef]

106. Weiss, M.; Dullenkopf, A.; Kolarova, A.; Schulz, G.; Frey, B.; Baenziger, O. Near-Infrared Spectroscopic Cerebral Oxygenation Reading in Neonates and Infants Is Associated with Central Venous Oxygen Saturation. *Paediatr. Anaesth.* **2005**, *15*, 102–109. [CrossRef] [PubMed]

107. Greisen, G.; Leung, T.; Wolf, M. Has the Time Come to Use Near-Infrared Spectroscopy as a Routine Clinical Tool in Preterm Infants Undergoing Intensive Care? *Philos. Trans. A Math. Phys. Eng. Sci.* **2011**, *369*, 4440–4451. [CrossRef]

108. van Bel, F.; Lemmers, P.; Naulaers, G. Monitoring Neonatal Regional Cerebral Oxygen Saturation in Clinical Practice: Value and Pitfalls. *Neonatology* **2008**, *94*, 237–244. [CrossRef]

109. Petrova, A.; Mehta, R. Near-Infrared Spectroscopy in the Detection of Regional Tissue Oxygenation during Hypoxic Events in Preterm Infants Undergoing Critical Care. *Pediatr. Crit. Care Med.* **2006**, *7*, 449. [CrossRef]

110. Edwards, A.D.; Richardson, C.; Cope, M.; Wyatt, J.S.; Delpy, D.T.; Reynolds, E.O.R. Cotside measurement of cerebral blood flow in ill newborn infants by near infrared spectroscopy. *Lancet* **1988**, *332*, 770–771. [CrossRef]
111. Cimatti, A.G.; Martini, S.; Galletti, S.; Vitali, F.; Aceti, A.; Frabboni, G.; Faldella, G.; Corvaglia, L. Cerebral Oxygenation and Autoregulation in Very Preterm Infants Developing IVH During the Transitional Period: A Pilot Study. *Front. Pediatr.* **2020**, *8*, 381. [CrossRef] [PubMed]
112. Baik, N.; Urlesberger, B.; Schwaberger, B.; Schmölzer, G.M.; Avian, A.; Pichler, G. Cerebral Haemorrhage in Preterm Neonates: Does Cerebral Regional Oxygen Saturation during the Immediate Transition Matter? *Arch. Dis. Child. Fetal. Neonatal. Ed.* **2015**, *100*, F422–F427. [CrossRef]
113. Hyttel-Sorensen, S.; Pellicer, A.; Alderliesten, T.; Austin, T.; van Bel, F.; Benders, M.; Claris, O.; Dempsey, E.; Franz, A.R.; Fumagalli, M.; et al. Cerebral near Infrared Spectroscopy Oximetry in Extremely Preterm Infants: Phase II Randomised Clinical Trial. *BMJ* **2015**, *350*, g7635. [CrossRef] [PubMed]
114. Plomgaard, A.M.; Alderliesten, T.; van Bel, F.; Benders, M.; Claris, O.; Cordeiro, M.; Dempsey, E.; Fumagalli, M.; Gluud, C.; Hyttel-Sorensen, S.; et al. No Neurodevelopmental Benefit of Cerebral Oximetry in the First Randomised Trial (SafeBoosC II) in Preterm Infants during the First Days of Life. *Acta Paediatr.* **2019**, *108*, 275–281. [CrossRef] [PubMed]
115. Hyttel-Sorensen, S.; Greisen, G.; Als-Nielsen, B.; Gluud, C. Cerebral Near-Infrared Spectroscopy Monitoring for Prevention of Brain Injury in Very Preterm Infants. *Cochrane Database Syst. Rev.* **2017**, *9*, CD011506. [CrossRef]
116. Mintzer, J.P.; Moore, J.E. Regional Tissue Oxygenation Monitoring in the Neonatal Intensive Care Unit: Evidence for Clinical Strategies and Future Directions. *Pediatr Res.* **2019**, *86*, 296–304. [CrossRef] [PubMed]
117. Chock, V.Y.; Frymoyer, A.; Yeh, C.G.; Van Meurs, K.P. Renal Saturation and Acute Kidney Injury in Neonates with Hypoxic Ischemic Encephalopathy Undergoing Therapeutic Hypothermia. *J. Pediatr.* **2018**, *200*, 232–239.e1. [CrossRef]
118. Ruf, B.; Bonelli, V.; Balling, G.; Hörer, J.; Nagdyman, N.; Braun, S.L.; Ewert, P.; Reiter, K. Intraoperative Renal Near-Infrared Spectroscopy Indicates Developing Acute Kidney Injury in Infants Undergoing Cardiac Surgery with Cardiopulmonary Bypass: A Case–Control Study. *Crit. Care* **2015**, *19*, 27. [CrossRef]
119. Vesoulis, Z.A.; Mintzer, J.P.; Chock, V.Y. Neonatal NIRS Monitoring: Recommendations for Data Capture and Review of Analytics. *J. Perinatol.* **2021**, *41*, 675–688. [CrossRef] [PubMed]
120. Ofman, G.; Tipple, T.E. Antioxidants & Bronchopulmonary Dysplasia: Beating the System or Beating a Dead Horse? *Free Radic. Biol. Med.* **2019**, *142*, 138–145. [CrossRef] [PubMed]
121. Boehme, J.; Le Moan, N.; Kameny, R.J.; Loucks, A.; Johengen, M.J.; Lesneski, A.L.; Gong, W.; Goudy, B.D.; Davis, T.; Tanaka, K.; et al. Preservation of Myocardial Contractility during Acute Hypoxia with OMX-CV, a Novel Oxygen Delivery Biotherapeutic. *PLoS Biol.* **2018**, *16*, e2005924. [CrossRef]
122. Tipple, T.E.; Welty, S.E.; Rogers, L.K.; Hansen, T.N.; Choi, Y.-E.; Kehrer, J.P.; Smith, C.V. Thioredoxin-Related Mechanisms in Hyperoxic Lung Injury in Mice. *Am. J. Respir. Cell Mol. Biol.* **2007**, *37*, 405–413. [CrossRef]
123. Li, Q.; Wall, S.B.; Ren, C.; Velten, M.; Hill, C.L.; Locy, M.L.; Rogers, L.K.; Tipple, T.E. Thioredoxin Reductase Inhibition Attenuates Neonatal Hyperoxic Lung Injury and Enhances Nuclear Factor E2-Related Factor 2 Activation. *Am. J. Respir. Cell Mol. Biol.* **2016**, *55*, 419–428. [CrossRef]
124. Dunigan, K.; Li, Q.; Li, R.; Locy, M.L.; Wall, S.; Tipple, T.E. The Thioredoxin Reductase Inhibitor Auranofin Induces Heme Oxygenase-1 in Lung Epithelial Cells via Nrf2-Dependent Mechanisms. *Am. J. Physiol. -Lung Cell Mol. Physiol.* **2018**, *315*, L545–L552. [CrossRef]
125. Ma, Q. Role of Nrf2 in Oxidative Stress and Toxicity. *Annu. Rev. Pharmacol. Toxicol.* **2013**, *53*, 401–426. [CrossRef]
126. Reiter, R.J.; Tan, D.X.; Osuna, C.; Gitto, E. Actions of Melatonin in the Reduction of Oxidative Stress. A Review. *J. Biomed. Sci.* **2000**, *7*, 444–458. [CrossRef]
127. Pan, L.; Fu, J.-H.; Xue, X.-D.; Xu, W.; Zhou, P.; Wei, B. Melatonin Protects against Oxidative Damage in a Neonatal Rat Model of Bronchopulmonary Dysplasia. *World J. Pediatr.* **2009**, *5*, 216–221. [CrossRef] [PubMed]
128. Robertson, N.J.; Martinello, K.; Lingam, I.; Avdic-Belltheus, A.; Meehan, C.; Alonso-Alconada, D.; Ragab, S.; Bainbridge, A.; Sokolska, M.; Tachrount, M.; et al. Melatonin as an Adjunct to Therapeutic Hypothermia in a Piglet Model of Neonatal Encephalopathy: A Translational Study. *Neurobiol. Dis.* **2019**, *121*, 240–251. [CrossRef] [PubMed]
129. Sampath, V.; Garland, J.S.; Helbling, D.; Dimmock, D.; Mulrooney, N.P.; Simpson, P.M.; Murray, J.C.; Dagle, J.M. Antioxidant Response Genes Sequence Variants and BPD Susceptibility in VLBW Infants. *Pediatr. Res.* **2015**, *77*, 477–483. [CrossRef] [PubMed]

Systematic Review

Male Disadvantage in Oxidative Stress-Associated Complications of Prematurity: A Systematic Review, Meta-Analysis and Meta-Regression

Elke van Westering-Kroon [1], Maurice J Huizing [1], Eduardo Villamor-Martínez [1,2] and Eduardo Villamor [1,*]

1 Department of Pediatrics, Maastricht University Medical Center (MUMC+), School for Oncology and Developmental Biology (GROW), 6229HA Maastricht, The Netherlands; elke.kroon@mumc.nl (E.v.W.-K.); m.huizing@mumc.nl (M.J.H.); e.villamorm@gmail.com (E.V.-M.)
2 Statistics Netherlands, 6412HX Heerlen, The Netherlands
* Correspondence: e.villamor@mumc.nl

Abstract: A widely accepted concept is that boys are more susceptible than girls to oxidative stress-related complications of prematurity, including bronchopulmonary dysplasia (BPD), retinopathy of prematurity (ROP), necrotizing enterocolitis (NEC), intraventricular hemorrhage (IVH), and periventricular leukomalacia (PVL). We aimed to quantify the effect size of this male disadvantage by performing a systematic review and meta-analysis of cohort studies exploring the association between sex and complications of prematurity. Risk ratios (RRs) and 95% CIs were calculated by a random-effects model. Of 1365 potentially relevant studies, 41 met the inclusion criteria (625,680 infants). Male sex was associated with decreased risk of hypertensive disorders of pregnancy, fetal distress, and C-section, but increased risk of low Apgar score, intubation at birth, respiratory distress, surfactant use, pneumothorax, postnatal steroids, late onset sepsis, any NEC, NEC > stage 1 (RR 1.12, CI 1.06–1.18), any IVH, severe IVH (RR 1.28, CI 1.22–1.34), severe IVH or PVL, any BPD, moderate/severe BPD (RR 1.23, CI 1.18–1.27), severe ROP (RR 1.14, CI 1.07–1.22), and mortality (RR 1.23, CI 1.16–1.30). In conclusion, preterm boys have higher clinical instability and greater need for invasive interventions than preterm girls. This leads to a male disadvantage in mortality and short-term complications of prematurity.

Keywords: preterm birth; oxidative stress; sex differences; male disadvantage; female advantage; bronchopulmonary dysplasia; retinopathy of prematurity; necrotizing enterocolitis; intraventricular hemorrhage; periventricular leukomalacia; mortality

Citation: van Westering-Kroon, E.; Huizing, M.J.; Villamor-Martínez, E.; Villamor, E. Male Disadvantage in Oxidative Stress-Associated Complications of Prematurity: A Systematic Review, Meta-Analysis and Meta-Regression. *Antioxidants* **2021**, *10*, 1490. https://doi.org/10.3390/antiox10091490

Academic Editors: Julia Kuligowski and Máximo Vento

Received: 30 August 2021
Accepted: 14 September 2021
Published: 18 September 2021

Publisher's Note: MDPI stays neutral with regard to jurisdictional claims in published maps and institutional affiliations.

1. Introduction

Preterm birth is defined as birth before 37 completed weeks of gestational age (GA) and is further subdivided in extremely (GA < 28 weeks), very (GA 28 to <32 weeks), moderate (GA 32 to <34 weeks), and late (GA 34 to <37 weeks) preterm birth. Prematurity, particularly in the in the lowest ranges of GA, is a leading cause of infant mortality, as well as long-term morbidity [1].

Two widely accepted concepts in neonatal medicine are the so-called "male disadvantage" and "oxygen radical disease in neonatology". The first concept is supported by a large body of evidence showing that boys are more susceptible than girls to adverse outcomes of prematurity, including bronchopulmonary dysplasia (BPD), retinopathy of prematurity (ROP), necrotizing enterocolitis (NEC), intraventricular hemorrhage (IVH), periventricular leukomalacia (PVL), chronic neurodevelopmental and cognitive impairment, and death [2–7].

The term "oxygen radical disease in neonatology" was coined by Saugstad in the 1980s when he hypothesized that complications of prematurity, such as BPD, ROP, NEC, IVH, or PVL, are different facets of one disease sharing a basic pathogenetic mechanism: increased oxidative stress and reduced endogenous antioxidant defenses [8,9]. Interestingly, preterm girls have higher antioxidant enzyme activity than preterm boys and it has been suggested that these differences may play a key role in the male disadvantage of prematurity [10–12].

Although the notion of male disadvantage of prematurity is more than five decades old, only sex-associated differences in mortality have been systematically reviewed [13]. Our current aim is to conduct a systematic review and meta-analysis on male-female differences in risk of developing oxidative stress-associated complications of prematurity. In addition to outcomes such as BPD, ROP, NEC, IVH, or PVL, we also analyzed potential male–female differences in obstetrical characteristics and clinical conditions in the first weeks of postnatal life. Finally, since it has been suggested that the male disadvantage of prematurity has undergone changes in the last few years [14], we investigated by meta-regression the influence of time and other variables on the association between infant sex and complications of prematurity.

2. Materials and Methods

The study was performed and reported according to the preferred reporting items for systematic reviews and meta-analyses (PRISMA) and meta-analysis of observational studies in epidemiology (MOOSE) guidelines [15]. Review protocol was registered in the PROSPERO international register of systematic reviews (ID = CRD42018095509). The research question was "Do preterm boys have a higher risk of developing short-term complications of prematurity than preterm girls?"

2.1. Sources and Search Strategy

A comprehensive literature search was undertaken using the PubMed and EMBASE databases. The search strategy is detailed in Table S1. No language limit was applied. The literature search was updated up to February 2021. Narrative reviews, systematic reviews, case reports, letters, editorials, and commentaries were excluded, but read to identify potential additional studies. Additional strategies to identify studies included manual review of reference lists from key articles that fulfilled our eligibility criteria, use of "related articles" feature in PubMed, and use of the "cited by" tool in Web of Science and Google scholar.

2.2. Study Selection

Studies were included if they had a prospective or retrospective cohort design, examined preterm (GA < 37 weeks) infants, and reported primary data that could be used to measure the association between infant sex and short-term complications of prematurity. We only selected studies in which infant sex was the independent variable and the perinatal characteristics and outcomes were the dependent variables. Studies that exclusively included late preterm infants (GA $\geq$ 34 weeks), or combined preterm and term infants were excluded. To identify relevant studies, two reviewers (E.V., E.V.-M.) independently screened the results of the searches and applied inclusion criteria using a structured form. Discrepancies were resolved through discussion or consultation with a third reviewer (M.J.H.).

2.3. Data Extraction and Quality Assessment

Two reviewers (E.V., E.V.-M.) extracted data from relevant studies using a predetermined data extraction form, and two reviewers (E.W.-K., M.J.H.) checked data extraction for accuracy and completeness. Discrepancies were resolved by consulting the primary report. Data extracted from each study included citation information, language of publication, study design, location and frame time, patient characteristics, and results (including raw numbers or summary statistics when raw numbers were not available). Data were extracted for all obstetric and perinatal variables as well as clinical conditions and outcomes reported in each study.

Methodological quality was assessed using the Newcastle-Ottawa Scale (NOS) for cohort studies [15]. This scale assigns a maximum of 9 points (4 for selection, 2 for comparability, and 3 for outcome). NOS scores ≥ 7 were considered high-quality studies (low risk of bias), and scores of 5 to 6 denoted moderate quality (moderate risk of bias) [15].

2.4. Statistical Analysis

Meta-analysis was performed when at least three studies were identified that reported on the same variable or outcome measure. Studies were combined and analyzed using comprehensive meta-analysis V3.0 software (Biostat Inc., Englewood, NJ, USA). Due to anticipated heterogeneity, summary statistics were calculated with a random-effects model. This model accounts for variability between studies as well as within studies. For dichotomous outcomes, the risk ratio (RR) with 95% confidence interval (CI) was calculated. For continuous outcomes (example: GA), the mean difference (MD) with 95% CI was calculated. Statistical heterogeneity was assessed by Cochran's Q statistic and by the I^2 statistic. Potential sources of heterogeneity were assessed through subgroup analysis and/or random effects (method of moments) univariate meta-regression analysis as previously described [16,17]. For both categorical and continuous covariates, the R^2 analog, defined as the total between-study variance explained by the moderator, was calculated based on the meta-regression matrix. Predefined sources of heterogeneity included the following characteristics of cohorts: mean or median GA, median year of birth, and geographical location (continent). We used the Egger's regression test and funnel plots to assess publication bias. Subgroup analyses, meta-regression, and publication bias assessment were performed only for the main outcomes (BPD, IVH, PVL, ROP, NEC, and mortality) and when there were at least ten studies in the meta-analysis. A probability value of less than 0.05 (0.10 for heterogeneity) was considered statistically significant.

3. Results

3.1. Description of Studies and Quality Assessment

The flow diagram of the search process is shown in Figure S1. Of 1365 potentially relevant studies, 41 (including 625,680 infants, 319,470 males) were included [2–5,14,18–53]. Their characteristics are summarized in Table S2. The percentage of males in the cohorts ranged from 41.2% [32] to 66.2% [20] with a pooled percentage of 52.2% (95% CI 51.4 to 53.0). Four studies included exclusively twin infants [24,35,44,51] and two studies included exclusively singleton infants [43,49]. The quality score of each study according to the Newcastle-Ottawa Scale is depicted in Table S2. All studies received at least seven points, indicating a low risk of bias.

3.1.1. Meta-Analysis

The following variables were reported in more than two studies and were therefore included in the meta-analysis: chorioamnionitis, hypertensive disorders of pregnancy, maternal diabetes, prenatal care, premature rupture of membranes, prolonged rupture of membranes, antepartum hemorrhage, antenatal corticosteroids, fetal distress, cesarean-section, birth in a non-tertiary hospital (outborn), 5′ Apgar score < 3, 5′ Apgar score < 7, intubation at birth, resuscitation at birth, birth weight (BW) below the 10th percentile, BW below the 3rd percentile or -2SD, early onset (<72 h) sepsis, late onset (>72 h) sepsis, undefined onset sepsis, hypotension, patent ductus arteriosus (PDA), respiratory distress syndrome (RDS), administration of surfactant, mechanical ventilation, pneumothorax, postnatal steroids, any BPD (defined as oxygen requirement on postnatal day 28), moderate/severe BPD (defined as oxygen requirement at the postmenstrual age of 36 weeks), any IVH (grade 1–4), severe IVH (grade 3–4), PVL, severe IVH or PVL, any ROP, severe ROP (stage $\geq$ 3 or requiring treatment), any NEC, NEC $\geq$ Stage II, and mortality before discharge.

The meta-analyses on obstetric and perinatal characteristics are summarized in Figure 1 and Table 1. The meta-analyses on clinical characteristics and outcomes are summarized in Figure 2 and Table 2. The individual meta-analyses for each outcome (BPD, IVH, PVL, ROP, NEC, and mortality) are shown in Figures S2–S12. Male sex was associated with a decreased risk of hypertensive disorders of pregnancy, fetal distress, and cesarean-section, but an increased risk of birth in a non-tertiary hospital, 5′ Apgar score < 3, intubation at birth, respiratory distress syndrome, surfactant use, pneumothorax, postnatal steroids, late onset sepsis, any BPD (Figures 2 and S2), moderate/severe BPD (Figures 2 and S3), any IVH (Figures 2 and S4), severe IVH (Figures 2 and S5), severe IVH or PVL (Figures 2 and S7), severe ROP (Figures 2 and S9), any NEC (Figures 2 and S10), NEC ≥ stage II (Figures 2 and S11), and mortality (Figures 2 and S12). With regard to the continuous variables, BW was significantly higher in boys than in girls (Table 1). In contrast, no differences were found by infant sex in either GA or maternal age (Table 1).

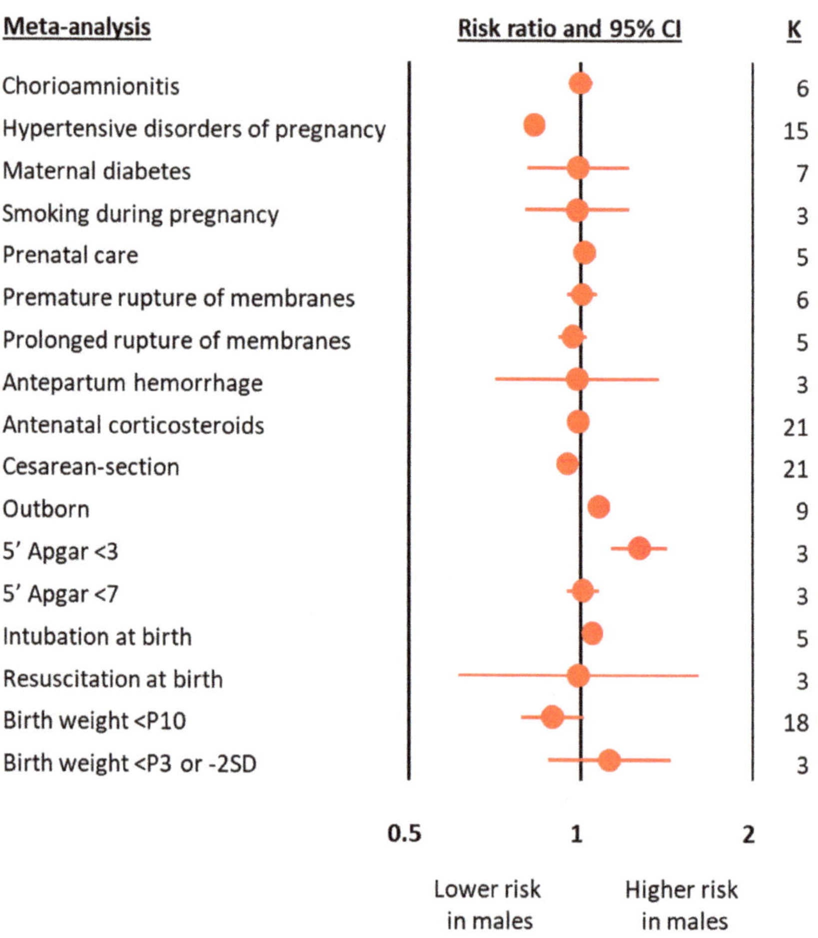

Figure 1. Summary of meta-analyses on the association between obstetric and perinatal characteristics of preterm infants and male sex. CI: confidence interval; K: number of studies; P3: 3rd percentile; P10: 10th percentile; SD: standard deviation.

Table 1. Meta-analyses on the association between obstetric and perinatal characteristics of preterm infants and male sex.

Meta-Analysis	K	RR	95% CI		p	Heterogeneity	
			Lower Limit	Upper Limit		I^2 (%)	p
Chorioamnionitis	6	1.001	0.953	1.052	0.969	67.3	0.009
Hypertensive disorders of pregnancy	15	0.829	0.803	0.856	<0.001	22.9	0.200
Maternal diabetes	7	0.991	0.808	1.214	0.927	65.9	0.007
Smoking during pregnancy	3	0.987	0.800	1.218	0.901	3.9	<0.001
Prenatal care	5	1.017	0.982	1.053	0.352	96.2	0.353
Premature rupture of membranes	6	1.006	0.947	1.068	0.852	62.8	0.020
Prolonged rupture of membranes	5	0.968	0.914	1.026	0.275	0.0	0.972
Antepartum hemorrhage	3	0.986	0.708	1.374	0.936	96.9	<0.001
Antenatal corticosteroids	21	0.992	0.982	1.003	0.143	44.9	0.012
Fetal distress	3	0.784	0.678	0.907	0.001	0.0	0.741
Cesarean-section	21	0.980	0.966	0.995	0.008	51.5	0.003
Outborn	9	1.077	1.027	1.128	0.002	0.0	0.682
Apgar 5′ <3	3	1.269	1.132	1.422	<0.001	0.0	0.726
Apgar 5′ <7	3	1.010	0.946	1.077	0.772	85.2	0.001
Intubation at birth	5	1.038	1.006	1.071	0.019	66.4	0.018
Resuscitation at birth	3	0.990	0.609	1.609	0.968	93.2	<0.001
Birth weight <P10	18	0.892	0.785	1.014	0.080	80.9	<0.001
Birth weight <P3	3	1.123	0.877	1.438	0.358	51.5	0.127
Continuous variables		**MD**					
Gestational age (weeks)	24	−0.10	−0.21	0.01	0.076	87.0	<0.001
Birth weight (g)	24	47.8	34.1	61.5	<0.001	91.5	<0.001
Maternal age (years)	10	0.0	−0.5	0.5	0.999	92.5	<0.001

Random effects analysis. Risk ratio (RR) > 1 indicates association of male sex with increased risk of the variable and RR < 1 indicates association of male sex with decreased risk of the variable. K: number of studies, MD: difference of means.

Neither visual inspection of funnel plots (Figure S13) nor Egger's test suggested publication or selection bias for any of the eligible meta-analyses (i.e., with at least ten studies).

3.1.2. Subgroup Analysis and Meta-Regression

Subgroup analysis based on the geographic location (continent) of the studies showed no significant differences for any of the outcomes analyzed, with only the exception of PVL (Table S2). The effect size of the association between male sex and PVL was significantly lower (meta-regression p = 0.048, R^2-analog = 0.5) in the cohorts from America when compared with Asian and European cohorts (Table S3).

Meta-regression showed that the effect size of the association between male sex and mortality significantly decreased as the median year of the cohort increased (Figure 3A). In contrast, the association between male sex and the other outcomes did not correlate with the median year of birth of the cohort (Table S4). Meta-regression also showed that the effect size of the association between male sex and mortality significantly increased as the mean/median gestational age of the cohort increased (Figure 3B). The association between sex and the other outcomes did not correlate with the mean/median gestational age of the cohort (Table S4).

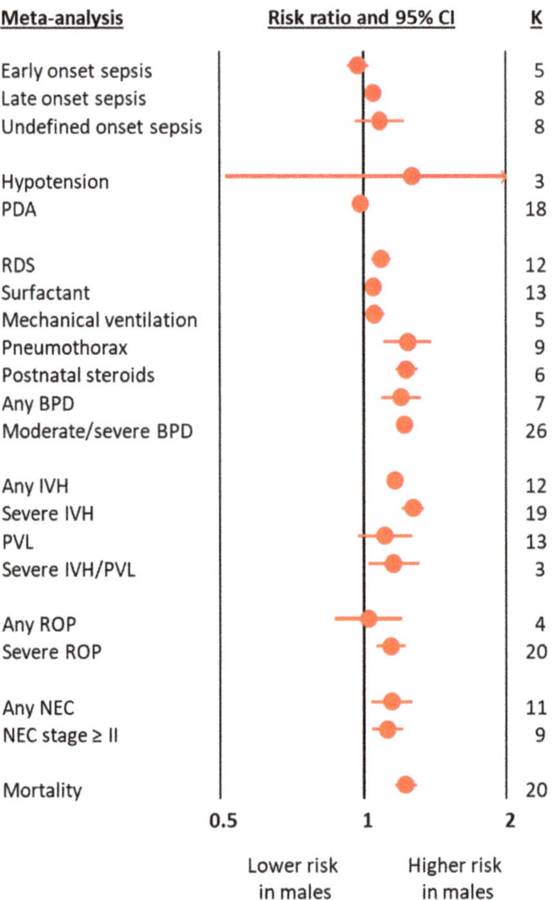

Figure 2. Summary of meta-analyses on the association between clinical characteristics and outcomes of preterm infants and male sex. BPD: bronchopulmonary dysplasia; CI: confidence interval; IVH: intraventricular hemorrhage; K: number of studies; NEC: necrotizing enterocolitis; PDA: patent ductus arteriosus; PVL: periventricular leukomalacia; ROP: retinopathy of prematurity.

Table 2. Meta-analyses on the association between clinical characteristics and outcomes of preterm infants and male sex.

Meta-Analysis	K	RR	95% CI Lower Limit	95% CI Upper Limit	*p*	Heterogeneity I^2 (%)	Heterogeneity *p*
Early onset sepsis	5	0.975	0.924	1.030	0.371	0.0	0.459
Late onset sepsis	8	1.051	1.026	1.077	<0.001	17.0	0.296
Undefined onset sepsis	8	1.083	0.962	1.218	0.186	20.6	0.266
Hypotension	3	1.270	0.514	3.140	0.605	72.7	0.026
PDA	18	0.985	0.958	1.012	0.262	52.5	0.004
RDS	12	1.090	1.042	1.140	<0.001	96.1	<0.001
Surfactant	13	1.031	1.026	1.036	<0.001	41.4	0.059
Mechanical ventilation	5	1.054	1.003	1.108	0.038	54.7	0.066

Table 2. *Cont.*

Pneumothorax	9	1.240	1.104	1.393	<0.001	42.2	0.086
Postnatal steroids	6	1.234	1.169	1.302	<0.001	37.6	0.433
Any BPD	7	1.200	1.091	1.319	<0.001	66.0	0.004
Moderate/severe BPD	26	1.219	1.176	1.264	<0.001	71.4	<0.001
Any IVH	12	1.166	1.139	1.193	<0.001	0.0	0.680
Severe IVH	19	1.271	1.207	1.338	<0.001	40.5	0.035
PVL	13	1.110	0.971	1.269	0.128	77.3	<0.001
Severe IVH/PVL	3	1.158	1.023	1.310	0.020	80.8	0.005
Any ROP	4	1.025	0.870	1.207	0.767	66.3	0.031
Severe ROP	20	1.143	1.065	1.226	<0.001	79.2	<0.001
Any NEC	11	1.145	1.036	1.266	0.008	60.1	0.003
NEC stage ≥ II	9	1.122	1.039	1.211	0.003	32.9	0.155
Mortality	20	1.227	1.163	1.294	<0.001	83.7	<0.001

Random effects analysis. Risk ratio (RR) > 1 indicates association of male sex with increased risk of the outcome and RR < 1 indicates association of male sex with decreased risk of the outcome. K: number of studies.

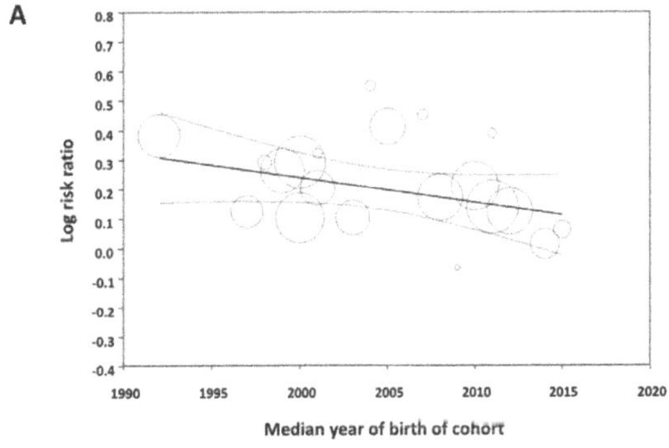

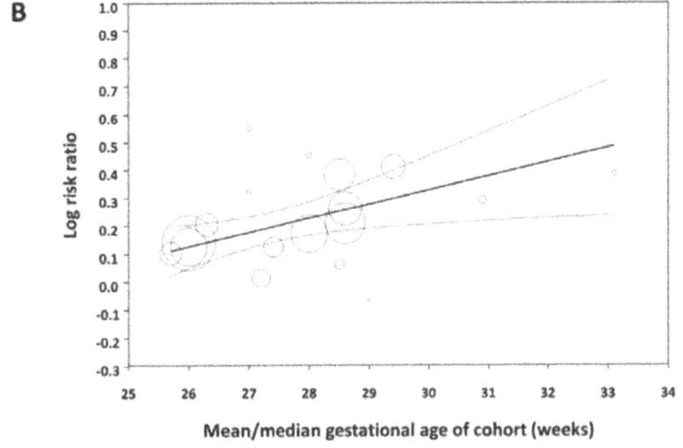

Figure 3. Meta-regression. (**A**) Plot showing the correlation between the association of male sex with mortality in preterm infants and the median year of birth of each cohort. A total of 24 studies were included (coefficient, −0.009; standard error, 0.004; $p = 0.019$; R^2-analog, 0.37). (**B**) Plot showing the correlation between the association of male sex with mortality in preterm infants and mean/median gestational age of each cohort. A total of 18 studies were included (coefficient, 0.051; standard error, 0.014; $p < 0.001$; R^2-analog, 0.75).

4. Discussion

To the best of our knowledge, this is the first systematic review and meta-analysis focused on male–female differences in short-term outcomes of prematurity. Our results confirm the presence of male disadvantage in mortality as well as relevant morbidities, including IVH, BPD, ROP, and NEC. Although the observed increases in risk are modest, they hold important implications for understanding preterm birth complications. Besides the short-term complications, we investigated whether other prognostic factors such as GA, birth weight, obstetric history, or clinical condition in the first days of life were different between boys and girls. We found no male–female differences in GA but the well-known difference is birth weight. In addition, meta-analysis showed that male sex was associated with decreased risk of being exposed to hypertension during pregnancy, developing fetal distress, and being born by cesarean section but increased risk of birth in a non-tertiary hospital, low Apgar score, intubation at birth, developing respiratory distress, being treated with surfactant and mechanical ventilation, developing pneumothorax, receiving postnatal steroids, and developing late onset sepsis. These differences in clinical course may have a major influence on the development of the pulmonary, neurological, ocular, and gastrointestinal complications of prematurity. On the other hand, our meta-analysis could not demonstrate that the rates of hypotension, PDA, or early onset sepsis were significantly different between boys and girls.

Male–female differences in human health and disease have been recognized for many years [54–57]. In reproductive and perinatal medicine, there are numerous studies dealing with sex differences, extending from fertilization and embryo implantation to the neonatal period [58–66]. The male to female ratio at birth is generally estimated to be around 1.05–1.06 [63–65,67]. This excess of males at birth is known for centuries and has been extensively studied by demographers, statisticians, epidemiologists, and biologists [63]. Focusing exclusively on preterm birth, the excess of males is even higher with male to female ratios around 1.2 [63–65,68,69]. However, the underlying mechanisms for this difference remain unclear, with suggestions including sexual dimorphism in embryonic and fetal homeostasis, as well as in the pathophysiological pathways that trigger preterm birth.

Preterm birth is always the result of a pathologic process, which may not only contribute to early delivery but may also adversely affect neonatal outcomes [16,70–73]. The pathophysiological pathways, or endotypes, leading to very and extreme preterm birth are divided into two main groups: (1) intrauterine infection/inflammation, and (2) dysfunctional placentation [16,70–73]. The first group is related to chorioamnionitis and placental microbial invasion and is associated with preterm labor, pre-labor premature rupture of membranes, placental abruption, and cervical insufficiency. The second group is associated with hypertensive disorders of pregnancy (including preeclampsia, eclampsia, and pregnancy-induced hypertension), and the entity identified as fetal indication/IUGR [16,70–73]. The association between fetal sex and prematurity endotype has been particularly examined in the case of the dysfunctional placentation endotype. A number of meta-analysis showed that preterm preeclampsia is associated with carrying a female fetus, while pregnancies with a male fetus are associated with developing term and post-term preeclampsia [61,62]. Accordingly, we observed an increased risk of hypertensive disorders of pregnancy and fetal distress associated with female sex. In contrast, there was no evidence of sexual dimorphism for conditions related to the infectious-inflammatory endotype, such as chorioamnionitis or rupture of membranes. Nevertheless, it should be taken into account that the low number of studies reporting on these prenatal conditions limits the power of the meta-analysis to detect possible differences. In fact, it has been suggested that pregnancy with a male fetus may favor a more pro-inflammatory intra-uterine environment, leading to a higher incidence of infection/inflammation-driven preterm birth [59,69,74]. Our group is currently conducting a meta-analysis exclusively focused on the association between fetal sex and endotype of prematurity.

Regardless of the imbalanced sex ratio at birth, the underlying mechanisms specifically responsible for the observed increase in neonatal morbidity in preterm boys are likely multifactorial and not yet fully elucidated. Possible explanations include male–female differences in mother–fetus interaction, rate of fetal development, molecular differences between sex chromosomes, epimutations that preferentially affect one sex, variations in antioxidant capacity, and hormonal differences [7,11,58–60,64–66]. Since the dawn of neonatology, the degree of pulmonary maturity at birth has been recognized as the critical factor in determining the survival and prognosis of preterm infants [75]. It was also noted early on that preterm boys had a higher rate and severity of respiratory distress syndrome than preterm girls [76]. This has been linked to the influence of sex hormones on lung development and maturation and to anatomic differences in lung development during fetal life [57,77,78]. As reviewed by Seaborn et al., the lung of male fetuses is exposed to higher levels of testosterone in the period preceding the surge of surfactant production [77,78]. Thereafter, both sexes are exposed to increasing levels of estradiol but the fetal lung has the capability to synthesize and inactivate sex hormones, and hence to modulate their action in a sex-dependent way [78]. The present meta-analysis confirms that the respiratory clinical course is less favorable in preterm boys than in preterm girls. As mentioned above, these differences in the first days of life may be critical for the later development of other complications of prematurity.

As mentioned in the introduction, oxidative stress plays a central pathogenic role in the development of most of the complications of prematurity [9,79,80]. Thus, sex differences in the development of antioxidant defenses are frequently pointed out as the key factor for male disadvantage among preterm infants [10–12]. From this perspective, the glutathione pathway is the most extensively studied [10–12]. Glutathione is the major endogenous soluble antioxidant in mammalian cells and its metabolism controls the intracellular levels of peroxides (via glutathione peroxidase), aldehydes (via glutathione S-transferase), and even radicals (via regeneration of oxidized vitamins C and E) [12]. As reviewed by Lavoie and Tremblay, numerous factors related to glutathione metabolism, including glutathione levels, activity of enzymes (glutathione peroxidase, glutathione reductase, glutathione S-transferase), and cellular uptake of cysteine have been found as sex-dependent in the placenta, umbilical cord, and blood cells of preterm infants [12]. Therefore, it has been suggested that antioxidant strategies in preterm infants should mainly target glutathione metabolism and be personalized considering, among others, the sex specificity [12].

The study cohorts included in our meta-analysis spanned a 30-year period (1986–2016) and some of the factors affecting outcome in the early 1990s may not be so relevant to current preterm populations. Moreover, it has been suggested that the advances in perinatal medicine, which have led to a decline in mortality and improved short-term outcomes for the most vulnerable preterm infants, have had a greater impact on boys than on girls [14]. Therefore, the male disadvantage of prematurity might be decreasing over the years [14]. We have tested this hypothesis by meta-regression and found that the male disadvantage in mortality among preterm infants tends to decrease as the cohorts include infants born in recent years. However, the increased risk of developing BPD, ROP, or NEC in males did not show this decreasing trend over the years. We also used meta-regression to analyze whether male disadvantage correlated with the gestational age of the cohort. Again, this meta-regression was only significant for mortality (Figure 3B). The association between male sex and risk of mortality decreased as the cohort had a lower mean gestational age. This effect of gestational age was not observed for any of the other complications of prematurity.

The major strength of our meta-analysis is the comprehensive database search to identify all the potential studies. Thus, the 41 included studies encompassed a total population of 625,680 infants from 16 different countries, providing a significant international representation. When we performed subgroup analyses based on continent, the only outcome where we found a significant geographic difference in sex ratio was PVL. It should be noted that, in contrast to the other outcomes analyzed, PVL was the only main complication of prematurity for which the meta-analysis did not show an increased risk associated with male sex. Subgroup analysis showed that the absence of male disadvantage for PVL was due to the marked differences between the American and the Asian and European cohorts (Table S2). Nevertheless, this finding may be an artifact due to the limited number of studies and therefore needs to be investigated in meta-analyses specifically focused on PVL. For all other outcomes, including IVH, BPD, ROP, NEC, and mortality, sub-group analysis did not show geographic differences, suggesting that the male disadvantage of prematurity is a ubiquitous phenomenon.

The main limitation of our systematic review is that studies were included only if sex was the independent variable and the association between sex and outcome was reported for more than one complication of prematurity. Although this design allowed for comparing the impact of male disadvantage on the different outcomes, we excluded a large number of studies in which an individual outcome was the independent variable and sex, among other potential risk factors, was the dependent variable. Our group is now analyzing these studies separately. The results of these meta-analyses, which for outcomes such as ROP or BPD include more than 250 studies, will confirm the present findings and analyze more comprehensively the influence of factors such as changes in trends over the years or geographic location on male disadvantage.

5. Conclusions

The present data suggest that the clinical course of preterm males is more complicated than that of females from the earliest moments of life. This higher clinical instability in males seems particularly to affect the respiratory system and leads to higher mortality and short-term morbidity. Complications such as BPD, ROP, NEC, IVH, or PVL will have a serious impact on post-discharge growth and neurodevelopment, extending the male disadvantage to the years of childhood and adolescence. In numerous studies on health conditions and neurocognitive outcomes of former preterm infants, adult females frequently perform better than adult males [1,7,81–83]. An improved understanding of sex-specific requirements of preterm infants may lead to optimized strategies to avoid the sequelae of early life oxidative stress and inflammation [7].

Supplementary Materials: The following are available online at https://www.mdpi.com/article/10.3390/antiox10091490/s1, Table S1: Search strategy; Table S2: Characteristics of the included studies; Table S3: Subgroup analyses based on continent; Table S4: Meta-regression analysis (continuous covariates); Figure S1: Flow diagram of the systematic search; Figure S2: Meta-analysis of the association between male sex of preterm infants and risk of bronchopulmonary dysplasia (BPD), defined as oxygen requirement on postnatal day 28; Figure S3: Meta-analysis of the association between male sex of preterm infants and risk of bronchopulmonary dysplasia (BPD), defined as oxygen requirement at the postmenstrual age of 36 weeks; Figure S4: Meta-analysis of the association between male sex and risk of intraventricular hemorrhage (IVH, grade 1–4) in preterm infants; Figure S5: Meta-analysis of the association between male sex and risk of severe intraventricular hemorrhage (IVH grade 3–4) in preterm infants; Figure S6: Meta-analysis of the association between male sex and risk of periventricular leukomalacia (PVL) in preterm infants; Figure S7: Meta-analysis of the association between male sex of preterm infants and risk of severe intraventricular hemorrhage (IVH grade 3–4) or periventricular leukomalacia (PVL); Figure S8: Meta-analysis of the association between male sex of preterm infants and risk of retinopathy of prematurity (ROP, any stage); Figure S9: Meta-analysis of the association between male sex in preterm infants and risk of severe retinopathy of prematurity (ROP grade ≥ 3 or requiring treatment); Figure S10: Meta-analysis of the association between male sex and risk of necrotizing enterocolitis (NEC, any stage) in preterm infants; Figure S11: Meta-analysis of the association between male sex and risk of necrotizing enterocolitis (NEC, stage ≥ II) in preterm infants; Figure S12: Meta-analysis of the association between male sex and risk of mortality before discharge in preterm infants; Figure S13: Funnel plot for publication bias analysis for the studies included in the different meta-analyses.

Author Contributions: Conceptualization, E.V. and E.V.-M.; methodology, E.V. and E.V.-M.; validation, E.v.W.-K., M.J.H..; formal analysis, E.V., E.v.W.-K., M.J.H. and E.V.-M.; investigation, E.V., E.v.W.-K., M.J.H. and E.V.-M.; data curation, E.V., E.v.W.-K., M.J.H. and E.V.-M.; writing—original draft preparation, E.V.; writing—review and editing, E.V., E.v.W.-K., M.J.H. and E.V.-M; visualization, E.V.; supervision, E.V. All authors have read and agreed to the published version of the manuscript.

Funding: This research received no external funding.

Institutional Review Board Statement: As this systematic review and meta-analysis did not involve animal subjects or personally identifiable information on human subjects, ethics review board approval and patient consent were not required.

Informed Consent Statement: Not applicable.

Data Availability Statement: All data relevant to the study are included in the article or uploaded as supplementary information. Additional data are available upon reasonable request.

Acknowledgments: We thank Zozaya for providing additional data from his study.

Conflicts of Interest: The authors declare no conflict of interest. The views expressed in this paper are those of the authors and do not necessarily reflect the policies of Statistics Netherlands.

References

1. Raju, T.N.; Buist, A.S.; Blaisdell, C.J.; Moxey-Mims, M.; Saigal, S. Adults born preterm: A review of general health and system-specific outcomes. *Acta Paediatr.* **2017**, *106*, 1409–1437. [CrossRef] [PubMed]
2. Ito, M.; Tamura, M.; Namba, F.; Neonatal Research Network of Japan. Role of sex in morbidity and mortality of very premature neonates. *Pediatr. Int.* **2017**, *59*, 898–905. [CrossRef]
3. Boghossian, N.S.; Geraci, M.; Edwards, E.M.; Horbar, J.D. Sex differences in mortality and morbidity of infants born at less than 30 weeks' gestation. *Pediatrics* **2018**, *142*, e20182352. [CrossRef] [PubMed]
4. Shim, S.-Y.; Cho, S.J.; Kong, K.A.; Park, E.A. Gestational age-specific sex difference in mortality and morbidities of preterm infants: A nationwide study. *Sci. Rep.* **2017**, *7*, 6161. [CrossRef]
5. Mohamed, M.A.; Aly, H. Male gender is associated with intraventricular hemorrhage. *Pediatrics* **2010**, *125*, e333–e339. [CrossRef] [PubMed]
6. O'Driscoll, D.N.; McGovern, M.; Greene, C.M.; Molloy, E.J. Gender disparities in preterm neonatal outcomes. *Acta Paediatr.* **2018**, *107*, 1494–1499. [CrossRef]
7. McDonald, F.B.; Dempsey, E.M.; O'Halloran, K.D. Caffeine therapy for apnoea of prematurity: Wake up to the fact that sex matters. *Exp. Physiol.* **2018**, *103*, 1294–1295. [CrossRef]
8. Saugstad, O.D. Hypoxanthine as an indicator of hypoxia: Its role in health and disease through free radical production. *Pediatr. Res.* **1988**, *23*, 143–150. [CrossRef]
9. Perez, M.; Robbins, M.E.; Revhaug, C.; Saugstad, O.D. Oxygen radical disease in the newborn, revisited: Oxidative stress and disease in the newborn period. *Free Rad. Biol. Med.* **2019**, *142*, 61–72. [CrossRef]
10. Vento, M.; Aguar, M.; Escobar, J.; Arduini, A.; Escrig, R.; Brugada, M.; Izquierdo, I.; Asensi, M.A.; Sastre, J.; Saenz, P. Antenatal steroids and antioxidant enzyme activity in preterm infants: Influence of gender and timing. *Antioxid. Redox Signal.* **2009**, *11*, 2945–2955. [CrossRef]
11. Lorente-Pozo, S.; Parra-Llorca, A.; Torres, B.; Torres-Cuevas, I.; Nuñez-Ramiro, A.; Cernada, M.; García-Robles, A.; Vento, M. Influence of sex on gestational complications, fetal-to-neonatal transition, and postnatal adaptation. *Front. Pediatr.* **2018**, *6*, 63. [CrossRef] [PubMed]
12. Lavoie, J.-C.; Tremblay, A. Sex-specificity of oxidative stress in newborns leading to a personalized antioxidant nutritive strategy. *Antioxidants* **2018**, *7*, 49. [CrossRef]
13. Vu, H.D.; Dickinson, C.; Kandasamy, Y. Sex difference in mortality for premature and low birth weight neonates: A systematic review. *Am. J. Perinatol.* **2018**, *35*, 707–715. [PubMed]
14. Garfinkle, J.; Yoon, E.W.; Alvaro, R.; Nwaesei, C.; Claveau, M.; Lee, S.K.; Shah, P.S. Trends in sex-specific differences in outcomes in extreme preterms: Progress or natural barriers? *Arch. Dis. Child. Fetal Neonatal Ed.* **2020**, *105*, 158–163. [CrossRef] [PubMed]
15. Stroup, D.F.; Berlin, J.A.; Morton, S.C.; Olkin, I.; Williamson, G.D.; Rennie, D.; Moher, D.; Becker, B.J.; Sipe, T.A.; Thacker, S.B. Meta-analysis of observational studies in epidemiology: A proposal for reporting. *JAMA* **2000**, *283*, 2008–2012. [CrossRef]
16. Wells, G.A.; Shea, B.; O'Connell, D.; Peterson, J.; Welch, V.; Losos, M.; Tugwell, P. The Newcastle-Ottawa Scale (NOS) for Assessing the Quality of Nonrandomised Studies in Meta-Analyses. Oxford. 2000. Available online: http://www.ohri.ca/programs/clinical_epidemiology/oxford.asp (accessed on 1 July 2020).
17. Pierro, M.; Villamor-Martinez, E.; van Westering-Kroon, E.; Alvarez-Fuente, M.; Abman, S.H.; Villamor, E. Association of the dysfunctional placentation endotype of prematurity with bronchopulmonary dysplasia: A systematic review, meta-analysis and meta-regression. *Thorax* **2021**. [CrossRef]
18. Villamor-Martinez, E.; Álvarez-Fuente, M.; Ghazi, A.M.; Degraeuwe, P.; Zimmermann, L.J.; Kramer, B.W.; Villamor, E. Association of chorioamnionitis with bronchopulmonary dysplasia among preterm infants: A systematic review, meta-analysis, and metaregression. *JAMA Netw. Open* **2019**, *2*, e1914611. [CrossRef]

19. Bertino, E.; Coscia, A.; Boni, L.; Rossi, C.; Martano, C.; Giuliani, F.; Fabris, C.; Spada, E.; Zolin, A.; Milani, S. Weight growth velocity of very low birth weight infants: Role of gender, gestational age and major morbidities. *Early Hum. Dev.* **2009**, *85*, 339–347. [CrossRef]
20. Binet, M.-E.; Bujold, E.; Lefebvre, F.; Tremblay, Y.; Piedboeuf, B.; Canadian Neonatal Network. Role of gender in morbidity and mortality of extremely premature neonates. *Am. J. Perinatol.* **2012**, *29*, 159–166. [CrossRef]
21. Chen, C.; Tian, T.; Liu, L.; Zhang, J.; Fu, H. Gender-related efficacy of pulmonary surfactant in infants with respiratory distress syndrome: A STROBE compliant study. *Medicine* **2018**, *97*, e0425. [CrossRef]
22. Derzbach, L.; Treszl, A.; Balogh, Á.; Vásárhelyi, B.; Tulassay, T. Gender dependent association between perinatal morbidity and estrogen receptor-alpha Pvull polymorphism. *J. Perinat. Med.* **2005**, *33*, 461–462. [CrossRef] [PubMed]
23. Deulofeut, R.; Dudell, G.; Sola, A. Treatment-by-gender effect when aiming to avoid hyperoxia in preterm infants in the NICU. *Acta Paediatr.* **2007**, *96*, 990–994. [CrossRef] [PubMed]
24. Elsmén, E.; Pupp, I.H.; Hellström-Westas, L. Preterm male infants need more initial respiratory and circulatory support than female infants. *Acta Paediatr.* **2004**, *93*, 529–533. [CrossRef] [PubMed]
25. Gagliardi, L.; Rusconi, F.; Reichman, B.; Adams, M.; Modi, N.; Lehtonen, L.; Kusuda, S.; Vento, M.; Darlow, B.A.; Bassler, D. Neonatal outcomes of extremely preterm twins by sex pairing: An international cohort study. *Arch. Dis. Child. Fetal Neonatal Ed.* **2021**, *106*, 17–24. [CrossRef]
26. Griesmaier, E.; Santuari, E.; Edlinger, M.; Neubauer, V.; Waltner-Romen, M.; Kiechl-Kohlendorfer, U. Differences in the maturation of amplitude-integrated EEG signals in male and female preterm infants. *Neonatology* **2014**, *105*, 175–181. [CrossRef]
27. Harris, C.; Zivanovic, S.; Lunt, A.; Calvert, S.; Bisquera, A.; Marlow, N.; Peacock, J.L.; Greenough, A. Lung function and respiratory outcomes in teenage boys and girls born very prematurely. *Pediatr. Pulmonol.* **2020**, *55*, 682–689. [CrossRef]
28. Hintz, S.R.; Kendrick, D.E.; Vohr, B.R.; Poole, W.K.; Higgins, R.D.; NICHD Neonatal Research Network. Gender differences in neurodevelopmental outcomes among extremely preterm, extremely-low-birthweight infants. *Acta Paediatr.* **2006**, *95*, 1239–1248. [CrossRef] [PubMed]
29. Huang, H.-M.; Lin, S.-A.; Chang, Y.-C.; Kuo, H.-K. Correlation between periventricular leukomalacia and retinopathy of prematurity. *Eur. J. Ophthalmol.* **2012**, *22*, 980–984. [CrossRef]
30. Jennische, M.; Sedin, G. Gender differences in outcome after neonatal intensive care: Speech and language skills are less influenced in boys than in girls at 6.5 years. *Acta Paediatr.* **2003**, *92*, 364–378. [CrossRef]
31. Jones, H.P.; Karuri, S.; Cronin, C.M.; Ohlsson, A.; Peliowski, A.; Synnes, A.; Lee, S.K. Actuarial survival of a large Canadian cohort of preterm infants. *BMC Pediatr.* **2005**, *5*, 40. [CrossRef]
32. Kent, A.L.; Wright, I.M.; Abdel-Latif, M.E. Mortality and adverse neurologic outcomes are greater in preterm male infants. *Pediatrics* **2012**, *129*, 124–131. [CrossRef]
33. Lauterbach, M.D.; Raz, S.; Sander, C.J. Neonatal hypoxic risk in preterm birth infants: The influence of sex and severity of respiratory distress on cognitive recovery. *Neuropsychology* **2001**, *15*, 411. [CrossRef] [PubMed]
34. Lavoie, M.E.; Robaey, P.; Stauder, J.E.; Glorieux, J.; Lefebvre, F. Extreme prematurity in healthy 5-year-old children: A re-analysis of sex effects on event-related brain activity. *Psychophysiology* **1998**, *35*, 679–689. [CrossRef] [PubMed]
35. Månsson, J.; Fellman, V.; Stjernqvist, K.; EXPRESS Study Group. Extremely preterm birth affects boys more and socio-economic and neonatal variables pose sex-specific risks. *Acta Paediatr.* **2015**, *104*, 514–521. [CrossRef] [PubMed]
36. Melamed, N.; Yogev, Y.; Glezerman, M. Effect of fetal sex on pregnancy outcome in twin pregnancies. *Obs. Gynecol.* **2009**, *114*, 1085–1092. [CrossRef] [PubMed]
37. Neubauer, V.; Griesmaier, E.; Ralser, E.; Kiechl-Kohlendorfer, U. The effect of sex on outcome of preterm infants—A population-based survey. *Acta Paediatr.* **2012**, *101*, 906–911. [CrossRef]
38. Peacock, J.L.; Marston, L.; Marlow, N.; Calvert, S.A.; Greenough, A. Neonatal and infant outcome in boys and girls born very prematurely. *Pediatr. Res.* **2012**, *71*, 305–310. [CrossRef]
39. Ramos-Navarro, C.; Sánchez-Luna, M.; Zeballos-Sarrato, S.; Pescador-Chamorro, I. Antenatal corticosteroids and the influence of sex on morbidity and mortality of preterm infants. *J. Matern. Fetal Neonatal Med.* **2020**, 1–8. [CrossRef]
40. Shinwell, E.S.; Reichman, B.; Lerner-Geva, L.; Boyko, V.; Blickstein, I. "Masculinizing" effect on respiratory morbidity in girls from unlike-sex preterm twins: A possible transchorionic paracrine effect. *Pediatrics* **2007**, *120*, e447–e453. [CrossRef]
41. Skiöld, B.; Alexandrou, G.; Padilla, N.; Blennow, M.; Vollmer, B.; Ådén, U. Sex differences in outcome and associations with neonatal brain morphology in extremely preterm children. *J. Pediatr.* **2014**, *164*, 1012–1018. [CrossRef]
42. Spinillo, A.; Montanari, L.; Gardella, B.; Roccio, M.; Stronati, M.; Fazzi, E. Infant sex, obstetric risk factors, and 2-year neurodevelopmental outcome among preterm infants. *Dev. Med. Child Neurol.* **2009**, *51*, 518–525. [CrossRef] [PubMed]
43. Stark, M.J.; Hodyl, N.A.; Wright, I.M.; Clifton, V. The influence of sex and antenatal betamethasone exposure on vasoconstrictors and the preterm microvasculature. *J. Matern. Fetal Neonatal Med.* **2011**, *24*, 1215–1220. [CrossRef] [PubMed]
44. Stark, M.J.; Wright, I.M.; Clifton, V.L. Sex-specific alterations in placental 11β-hydroxysteroid dehydrogenase 2 activity and early postnatal clinical course following antenatal betamethasone. *Am. J. Physiol. Regul. Integrat. Compar. Physiol.* **2009**, *297*, R510–R514. [CrossRef] [PubMed]
45. Steen, E.E.; Källén, K.; Maršál, K.; Norman, M.; Hellström-Westas, L. Impact of sex on perinatal mortality and morbidity in twins. *J. Perinat. Med.* **2014**, *42*, 225–231. [CrossRef] [PubMed]

46. Stevenson, D.K.; Verter, J.; Fanaroff, A.A.; Oh, W.; Ehrenkranz, R.A.; Shankaran, S.; Donovan, E.F.; Wright, L.L.; Lemons, J.A.; Tyson, J.E. Sex differences in outcomes of very low birthweight infants: The newborn male disadvantage. *Arch. Dis. Child. Fetal Neonatal Ed.* **2000**, *83*, F182–F185. [CrossRef] [PubMed]

47. Štimac, T.; Šopić-Rahelić, A.-M.; Ivandić, J.; Ekinja, E.; Blickstein, I. Effect of gender on growth-restricted fetuses born preterm. *J. Perinat. Med.* **2019**, *47*, 677–679. [CrossRef]

48. Tioseco, J.A.; Aly, H.; Essers, J.; Patel, K.; El-Mohandes, A.A. Male sex and intraventricular hemorrhage. *Pediatr. Crit. Care Med.* **2006**, *7*, 40–44. [CrossRef]

49. Tottman, A.C.; Bloomfield, F.H.; Cormack, B.E.; Harding, J.E.; Taylor, J.; Alsweiler, J.M. Sex-specific relationships between early nutrition and neurodevelopment in preterm infants. *Pediatr. Res.* **2020**, *87*, 872–878. [CrossRef]

50. Walker, M.; Fitzgerald, B.; Keating, S.; Ray, J.; Windrim, R.; Kingdom, J. Sex-specific basis of severe placental dysfunction leading to extreme preterm delivery. *Placenta* **2012**, *33*, 568–571. [CrossRef] [PubMed]

51. Wang, L.-W.; Lin, Y.-C.; Wang, S.-T.; Huang, C.-C.; Taiwan Premature Infant Developmental Collaborative Study Group. Identifying risk factors shared by bronchopulmonary dysplasia, severe retinopathy, and cystic periventricular leukomalacia in very preterm infants for targeted intervention. *Neonatology* **2018**, *114*, 17–24. [CrossRef] [PubMed]

52. Zhao, D.; Zou, L.; Lei, X.; Zhang, Y. Gender differences in infant mortality and neonatal morbidity in mixed-gender twins. *Sci. Rep.* **2017**, *7*, 8736. [CrossRef] [PubMed]

53. Zisk, J.L.; Genen, L.H.; Kirkby, S.; Webb, D.; Greenspan, J.; Dysart, K. Do premature female infants really do better than their male counterparts? *Am. J. Perinatol.* **2011**, *28*, 241–246. [CrossRef]

54. Zozaya, C.; Avila-Alvarez, A.; Arruza, L.; Rodrigo, F.G.-M.; Fernandez-Perez, C.; Castro, A.; Cuesta, M.T.; Vacas, B.; Couce, M.L.; Torres, M.V. The effect of morbidity and sex on postnatal growth of very preterm infants: A multicenter cohort study. *Neonatology* **2019**, *115*, 348–354. [CrossRef]

55. Editorial. Putting gender on the agenda. *Nature* **2010**, *465*, 665. [CrossRef] [PubMed]

56. Kardys, I.; Vliegenthart, R.; Oudkerk, M.; Hofman, A.; Witteman, J.C. The female advantage in cardiovascular disease: Do vascular beds contribute equally? *Am. J. Epidemiol.* **2007**, *166*, 403–412. [CrossRef] [PubMed]

57. Van Oyen, H.; Nusselder, W.; Jagger, C.; Kolip, P.; Cambois, E.; Robine, J.-M. Gender differences in healthy life years within the EU: An exploration of the "health–survival" paradox. *Int. J. Public Health* **2013**, *58*, 143–155. [CrossRef] [PubMed]

58. Townsend, E.A.; Miller, V.M.; Prakash, Y. Sex differences and sex steroids in lung health and disease. *Endocr. Rev.* **2012**, *33*, 1–47. [CrossRef] [PubMed]

59. Gabory, A.; Roseboom, T.; Moore, T.; Moore, L.; Junien, C. Placental contribution to the origins of sexual dimorphism in health and diseases: Sex chromosomes and epigenetics. *Biol. Sex Differ.* **2013**, *4*, 5. [CrossRef] [PubMed]

60. Clifton, V. Sex and the human placenta: Mediating differential strategies of fetal growth and survival. *Placenta* **2010**, *31*, S33–S39. [CrossRef]

61. Clifton, V.; Stark, M.; Osei-Kumah, A.; Hodyl, N. The feto-placental unit, pregnancy pathology and impact on long term maternal health. *Placenta* **2012**, *33*, S37–S41. [CrossRef]

62. Broere-Brown, Z.A.; Adank, M.C.; Benschop, L.; Tielemans, M.; Muka, T.; Gonçalves, R.; Bramer, W.M.; Schoufour, J.D.; Voortman, T.; Steegers, E.A. Fetal sex and maternal pregnancy outcomes: A systematic review and meta-analysis. *Biol. Sex Differ.* **2020**, *11*, 26. [CrossRef]

63. Schalekamp-Timmermans, S.; Arends, L.R.; Alsaker, E.; Chappell, L.; Hansson, S.; Harsem, N.K.; Jälmby, M.; Jeyabalan, A.; Laivuori, H. Fetal sex-specific differences in gestational age at delivery in pre-eclampsia: A meta-analysis. *Int. J. Epidemiol.* **2017**, *46*, 632–642.

64. James, W.H.; Grech, V. A review of the established and suspected causes of variations in human sex ratio at birth. *Early Hum. Dev.* **2017**, *109*, 50–56. [CrossRef]

65. Ingemarsson, I. Gender aspects of preterm birth. *BJOG Int. J. Obs. Gynaecol.* **2003**, *110*, 34–38. [CrossRef]

66. Cooperstock, M.; Campbell, J. Excess males in preterm birth: Interactions with gestational age, race, and multiple birth. *Obs. Gynecol.* **1996**, *88*, 189–193. [CrossRef]

67. DiPietro, J.A.; Voegtline, K.M. The gestational foundation of sex differences in development and vulnerability. *Neuroscience* **2017**, *342*, 4–20. [CrossRef] [PubMed]

68. Martin, J.A.; Hamilton, B.E.; Osterman, M.J.; Driscoll, A.K. Births: Final data for 2019. *Natl. Vital Stat. Rep.* **2021**, *70*, 1–51.

69. Peelen, M.J.; Kazemier, B.M.; Ravelli, A.C.; De Groot, C.J.; Van Der Post, J.A.; Mol, B.W.; Hajenius, P.J.; Kok, M. Impact of fetal gender on the risk of preterm birth, a national cohort study. *Acta Obs. Gynecol. Scand.* **2016**, *95*, 1034–1041. [CrossRef]

70. Challis, J.; Newnham, J.; Petraglia, F.; Yeganegi, M.; Bocking, A. Fetal sex and preterm birth. *Placenta* **2013**, *34*, 95–99. [CrossRef] [PubMed]

71. McElrath, T.F.; Hecht, J.L.; Dammann, O.; Boggess, K.; Onderdonk, A.; Markenson, G.; Harper, M.; Delpapa, E.; Allred, E.N.; Leviton, A. Pregnancy disorders that lead to delivery before the 28th week of gestation: An epidemiologic approach to classification. *Am. J. Epidemiol.* **2008**, *168*, 980–989. [CrossRef]

72. Gagliardi, L. Pregnancy complications and neonatal outcomes: Problems and perspectives. *Acta Paediatr.* **2014**, *103*, 682–683. [CrossRef] [PubMed]

73. Gagliardi, L.; Rusconi, F.; Da Fre, M.; Mello, G.; Carnielli, V.; Di Lallo, D.; Macagno, F.; Miniaci, S.; Corchia, C.; Cuttini, M. Pregnancy disorders leading to very preterm birth influence neonatal outcomes: Results of the population-based ACTION cohort study. *Pediatr. Res.* **2013**, *73*, 794–801. [CrossRef]

74. Gagliardi, L.; Rusconi, F.; Bellu, R.; Zanini, R.; Italian Neonatal Network. Association of maternal hypertension and chorioamnionitis with preterm outcomes. *Pediatrics* **2014**, *134*, e154–e161. [CrossRef]

75. Ghidini, A.; Salafia, C.M. Histologic placental lesions in women with recurrent preterm delivery. *Acta Obs. Gynecol. Scand.* **2005**, *84*, 547–550. [CrossRef] [PubMed]

76. Cataltepe, D. Historical perspectives: Beyond the first breath: Hyaline membrane disease and constructing the neonatal patient, 1959–1975. *NeoReviews* **2018**, *19*, e636–e644. [CrossRef]

77. Farrell, P.M.; Avery, M.E. Hyaline membrane disease. *Am. Rev. Respir. Dis.* **1975**, *111*, 657–688. [PubMed]

78. Raghavan, D.; Jain, R. Increasing awareness of sex differences in airway diseases. *Respirology* **2016**, *21*, 449–459. [CrossRef] [PubMed]

79. Seaborn, T.; Simard, M.; Provost, P.R.; Piedboeuf, B.; Tremblay, Y. Sex hormone metabolism in lung development and maturation. *Trends Endocrinol. Metabol.* **2010**, *21*, 729–738. [CrossRef]

80. Huizing, M.J.; Cavallaro, G.; Moonen, R.M.; González-Luis, G.E.; Mosca, F.; Vento, M.; Villamor, E. Is the C242T polymorphism of the CYBA gene linked with oxidative stress-associated complications of prematurity? *Antioxid. Redox Signal.* **2017**, *27*, 1432–1438. [CrossRef]

81. Linsell, L.; Malouf, R.; Morris, J.; Kurinczuk, J.J.; Marlow, N. Prognostic factors for poor cognitive development in children born very preterm or with very low birth weight: A systematic review. *JAMA Pediatr.* **2015**, *169*, 1162–1172. [CrossRef]

82. Linsell, L.; Johnson, S.; Wolke, D.; O'Reilly, H.; Morris, J.K.; Kurinczuk, J.J.; Marlow, N. Cognitive trajectories from infancy to early adulthood following birth before 26 weeks of gestation: A prospective, population-based cohort study. *Arch. Dis. Child.* **2018**, *103*, 363–370. [CrossRef] [PubMed]

83. Darlow, B.A.; Martin, J.; Horwood, L.J. Metabolic syndrome in very low birth weight young adults and controls: The New Zealand 1986 VLBW Study. *J. Pediatr.* **2019**, *206*, 128–133.e5. [CrossRef] [PubMed]

 *antioxidants* 

Review

The Effect of Pasteurization on the Antioxidant Properties of Human Milk: A Literature Review

Hannah G. Juncker [1,2,†], Eliza J. M. Ruhé [1,†], George L. Burchell [3], Chris H. P. van den Akker [4], Aniko Korosi [2], Johannes B. van Goudoever [1,*] and Britt J. van Keulen [1]

1   Amsterdam UMC, Emma Children's Hospital, Department of Pediatrics,
Amsterdam Reproduction & Development Research Institute, 1105 AZ Amsterdam, The Netherlands;
h.juncker@amsterdamumc.nl (H.G.J.); e.j.ruhe@amsterdamumc.nl (E.J.M.R.);
b.j.vankeulen@amsterdamumc.nl (B.J.v.K.)

2   Swammerdam Institute for Life Sciences—Center for Neuroscience, University of Amsterdam,
1098 XH Amsterdam, The Netherlands; A.Korosi@uva.nl

3   Medical Library, Vrije Universiteit Amsterdam, 1081 HV Amsterdam, The Netherlands; g.b.burchell@vu.nl

4   Amsterdam UMC, Emma Children's Hospital, Department of Pediatrics–Neonatology,
1105 AZ Amsterdam, The Netherlands; c.h.vandenakker@amsterdamumc.nl

*   Correspondence: h.vangoudoever@amsterdamumc.nl

†   These authors contributed equally to this work.

Abstract: High rates of oxidative stress are common in preterm born infants and have short- and long-term consequences. The antioxidant properties of human milk limits the consequences of excessive oxidative damage. However, as the mother's own milk it is not always available, donor milk may be provided as the best alternative. Donor milk needs to be pasteurized before use to ensure safety. Although pasteurization is necessary for safety reasons, it may affect the activity and concentration of several biological factors, including antioxidants. This literature review describes the effect of different pasteurization methods on antioxidant properties of human milk and aims to provide evidence to guide donor milk banks in choosing the best pasteurization method from an antioxidant perspective. The current literature suggests that Holder pasteurization reduces the antioxidant properties of human milk. Alternative pasteurization methods seem promising as less reduction is observed in several studies.

Keywords: donor milk; treatment; Holder pasteurization; oxidative stress; breastmilk; preterm; antioxidant capacity

Citation: Juncker, H.G.; Ruhé, E.J.M.; Burchell, G.L.; van den Akker, C.H.P.; Korosi, A.; van Goudoever, J.B.; van Keulen, B.J. The Effect of Pasteurization on the Antioxidant Properties of Human Milk: A Literature Review. *Antioxidants* **2021**, *10*, 1737. https://doi.org/10.3390/antiox10111737

Academic Editors: Julia Kuligowski and Máximo Vento

Received: 19 October 2021
Accepted: 29 October 2021
Published: 29 October 2021

Publisher's Note: MDPI stays neutral with regard to jurisdictional claims in published maps and institutional affiliations.

1. Introduction

Excessive oxidative stress, the release of ample reactive oxygen species (ROS) in reaction to a broad range of stressors, is common in preterm born infants. They are often exposed to stressors including infections, oxygen supplementation, phototherapy, and parenteral nutrition. ROS are essential at a certain level as they contribute considerably to biological homeostasis and play a role in cell signaling in psychological and pathophysiological processes but can also cause damage to molecules or even cells when present in excess [1]. Disproportionate ROS levels lead to irreversible cell damage, necrosis and apoptosis as a consequence of lipid peroxidation, protein alterations, and DNA oxidation [1]. Accumulation of ROS damage has short- and long-term consequences. As adequate concentrations of antioxidants are absent in preterm born infants, they are highly susceptible to damage from oxidative stress. Moreover, they have an impaired ability to increase the synthesis of antioxidants [2]. Subsequently, preterm infants have an increased risk of developing oxidative stress-related diseases, such as bronchopulmonary dysplasia, necrotizing enterocolitis, periventricular leukomalacia, and retinopathy of prematurity [3–5].

Human milk is the gold standard for early life nutrition, especially for preterm born infants. It is not only important for its nutritive value, but also contains numerous different classes of bioactive components, including antioxidant molecules. The antioxidant capacity of human milk refers to the total activity performed by all of the antioxidant molecules present in the milk. Indeed, fresh human milk has a better antioxidant profile than formula [6–9]. These antioxidant molecules can counteract the detrimental effects of oxidative stress [10]. Mothers who give birth to preterm infants often have a delayed initiation of lactation. Caesarean section, a very common delivery mode for preterm infants, is also associated with a reduction in breastfeeding rates. Consequently, human milk is often not sufficiently available despite the extreme importance of human milk during this stressful period early in life. When the mother's own milk is not available, donor milk is considered the best alternative [11].

As several viruses, such as human immunodeficiency virus, hepatitis B, cytomegalovirus and bacterial pathogens such as *Escherichia coli*, group B hemolytic streptococci, and other pathogens, could be transmitted via human milk, it is necessary to process donor milk before use to ensure it is safe for consumption by the recipient infant [12,13]. Pasteurization is an effective method to ensure the microbiological quality of the milk in human milk banks [14]. Several methods of pasteurization have been developed including Holder pasteurization (30 min at 62.5 °C), high-temperature short-time pasteurization (15 s at 72 °C), high-pressure treatment (usually 300–800 Megapascal (MPa), <5–10 min) [15]), and ultraviolet-C treatment (200–280 nm [15]). To date, Holder pasteurization is by far the most commonly used method in human milk banks [16]. Although pasteurization is necessary, it may change the composition of human milk. It is known that Holder pasteurization reduces important nutrients in human milk, for example amino acids and vitamins [17]. Moreover, it has been demonstrated that growth of preterm infants receiving pasteurized milk is lower compared to preterm infants receiving unpasteurized milk [18]. It can be hypothesized that pasteurization also changes the activity and concentration of several biological factors, including antioxidants. Currently a clear understanding on the effect of the various pasteurization methods influence human milk antioxidant properties is lacking. Therefore we set out to review the literature describing the effect of different pasteurization or treatment methods on the antioxidant properties of human milk and to provide evidence to guide donor milk banks in choosing the best pasteurization method.

2. Materials and Methods

2.1. Search Strategy

A systematic search was performed in the databases PubMed, Embase.com, Clarivate Analytics/Web of Science Core Collection, Cumulative Index to Nursing and Allied Health Literature (CINAHL), and the Wiley/Cochrane Library. The timeframe within the databases was from inception to 27 May 2021. The search included keywords and free text terms for (synonyms of) 'antioxidant' or 'oxidative stress' combined with (synonyms of) 'human milk' combined with (synonyms of) 'pasteurization'. A full overview of the search terms per database can be found in the Appendix A (see Tables A1–A5). No limitations on language were applied in the search. Abstracts and conference reports were not excluded.

2.2. Data Collection

As shown in Figure 1, 467 records were identified through database searching. After removal of duplicates, 307 titles and abstracts were screened. Twenty-six of these articles were eligible for full text screening, out of which 13 articles were included. Seven additional articles were identified through cross referencing. In total, 20 articles were included in the current literature review. Few articles reported on the same samples. Provided that extra information was presented, all articles were included in the review.

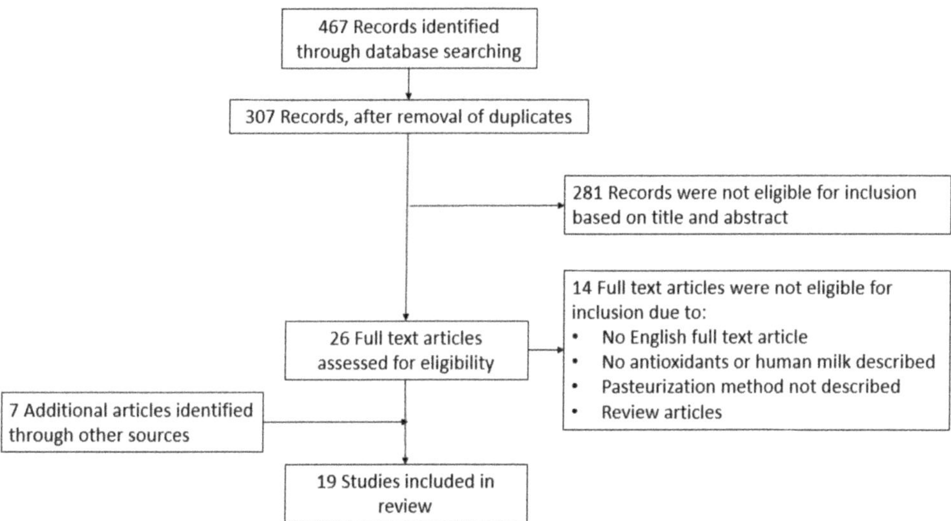

Figure 1. This flowchart presents the different phases of the review, according to the PRISMA-statement.

Antioxidant capacity was classified as follows (1) total antioxidant capacity, (2) enzymatic antioxidants, and (3) non-enzymatic antioxidants. All of the articles were sorted according to the categories above and were subsequently divided into the different pasteurization methods. Data were extracted from the articles and systematically summarized. If an article belonged to more than one category, the data were included in all applicable categories.

3. Results

3.1. Pasteurization Methods

The studies included in this review used several pasteurization methods. Thermal processing is the most commonly used pasteurization method, with Holder pasteurization as the gold standard described most frequently [6,13,14,19–32]. During Holder pasteurization, human milk is heated in a water bath to 62.5 °C for 30 min, and subsequently cooled to approximately 4 °C [33]. Some studies used alternative thermal heating at different temperatures for different durations, for example, high-temperature, short-time pasteurization (72–75 °C for at least 10 s), which was used in three articles [33].

Another pasteurization method is high-pressure treatment, which involves applying hydrostatic high pressure (300–800 MPa) to human milk for 5–10 min [20]. This method was used in four studies [19,20,28,29]. In three of the included articles, thermal and high pressure processing were combined [34].

Other alternative pasteurization methods were reported. (1) Ultraviolet-C treatment is a non-thermal pasteurization method using ultraviolet-c irradiation with a wavelength between 200–280 nm that is generally used in the food industry [21]. (2) During microwave processing, milk samples are treated at 2450 MHz, 800 W until a certain temperature is reached. Subsequently, the human milk samples are cooled [22]. (3) Thermosonication combines heat and ultrasound treatments. Samples are treated in an ultrasound bath at the frequency of 40 KHz and a power of 100 W [14]. The human milk samples are thermosonicated when the water temperature stabilizes (60 °C for 4 min). Subsequently, the samples are cooled to 5 °C [14]. No studies reported the effect of gamma-irradiation on the antioxidant capacity of human milk.

3.2. Effects of Pasteurization on the Antioxidant Properties of Human Milk

3.2.1. Total Antioxidant Capacity

Table 1 summarizes the results of the included studies that determined the effect of different pasteurization or treatment methods on the total antioxidant capacity. Total antioxidant capacity (TAC) is a general measure to indicate the level of free radicals scavenged by a test solution that is commonly used to assess the antioxidant status of human milk samples [35]. Different methods to determine the TAC are described in the literature such as: Trolox equivalent antioxidant capacity (TEAC), ferric reducing ability of plasma (FRAP), and cupric reducing antioxidant capacity (CUPRAC) [35]. In the included studies, the most frequently described method was the TEAC assay, which measures the TAC by scavenging 2,2-azinob-bis 3 ethylbenzthiazoline-6-sulfonic acid radicals (ABTS) [14,21,23–25]. Moreover, two studies used the DPPH assay, which measures the TAC by scavenging 1,1-diphenyl-2-picrylhydrazyl (DPPH) free radicals [14,23]. Furthermore, some studies used the FRAP method, which measures the extent of the reduction of ferric-tripyridyl triazine (Fe^{3+}-TPTZ) to ferrous tripyridyl triazine (Fe^{2+}-TPTZ) caused by the antioxidant in question [24].

Studies on the effect of pasteurization on the total antioxidant capacity of human milk show conflicting results. Some studies showed a reduction in total antioxidant capacity after Holder pasteurization [13,14,23] compared with untreated milk, while others showed no influence of Holder pasteurization [6,21,23–25].

Other pasteurization techniques, including high-pressure treatment, alternative thermal pasteurization, and ultraviolet-C treatment, did not influence the TAC [14,21,25,26]. However, Ramírez et al. investigated the combination of high-pressure treatment and thermal processing and found an increase in the total antioxidant capacity under the conditions of 400 MPa at $-15\,°C$ and at 800 MPa at $30\,°C$ [26]. An increase in total antioxidant capacity was also reported for thermosonication pasteurization (40 kHz, 100 W, and $60\,°C$ for 4 min).

3.2.2. Enzymatic Antioxidants

Table 2 summarizes the results of the included studies that determined the effect of different pasteurization or treatment methods on the enzymatic antioxidants.

Superoxide dismutase: superoxide dismutase is essential for the elimination of superoxide radicals in cells. It catalyzes the reaction of superoxide into hydrogen peroxide and dioxygen [36]. The produced hydrogen peroxide will subsequently be converted into water through other antioxidant enzymes such as catalase and glutathione peroxidase [36]. Marinković et al. showed that Holder pasteurization caused a decrease in superoxide dismutase activity [6]. Others showed a reduction following heating from $70\,°C$ upwards only [22]. High-pressure treatment increased the activity of superoxide dismutase by 57% at 193 MPa [25]. Microwave heating also led to a significant increase in superoxide dismutase activity (10%, 21%, and 34% at 62.5, 66, and $70\,°C$ for 1 min, respectively) [22].

Catalase: catalase is a dismutase enzyme as well and counters the detrimental action of free radicals in the cell. It plays an essential role in the degradation of hydrogen peroxide into water and oxygen [36]. Catalase activity was reduced by approximately 60% after Holder pasteurization [21,22]. Following alternative heat treatment ($70\,°C$ for 30 min), a 66% decrease in activity was observed [22]. Microwave heating at 62.5, 66, and $70\,°C$ for 1 min also reduced catalase activity in human milk by 34%, 42%, and 38%, respectively [22]. Ultraviolet-C treatment did not affect catalase activity [21].

Glutathione peroxidase: glutathione peroxidase plays an important role in catalyzing the decomposition of hydrogen peroxide in cells. Furthermore, glutathione peroxidase protects the cell against oxidative stress through the decomposition of lipid hydroperoxides [36]. Several studies have investigated the effect of pasteurization on glutathione peroxidase activity. Holder pasteurization showed a reduction in glutathione peroxidase in human milk, ranging from 23% to 70% [6,13,22,27]. Alternative heating methods showed a reduction from 41% to 56% [22]. After high-pressure treatment, glutathione peroxidase activity in human milk decreased by 62% [13]. Microwave heating for 1 min at temperatures of 62.5, 66, and 70 °C led to a 38%, 38%, and 53% decrease in glutathione peroxidase activity, respectively [22].

3.2.3. Non-Enzymatic Antioxidants

The antioxidant properties of human milk are also affected by non-enzymatic antioxidants, including glutathione, vitamin C, E, A, and other agents. Table 3 summarizes the results of the included studies that determined the effect of different pasteurization or treatment methods on the non-enzymatic antioxidants.

Glutathione: Glutathione is an important antioxidant in human milk, as it plays a role in the deactivation of oxygen-derived free radicals and the elimination of toxins, carcinogens, and malonic dialdehyde [37]. Silvestre, et al. investigated the effects of Holder pasteurization and high-temperature, short-time pasteurization, and found that Holder, but not HTST pasteurization reduced glutathione concentrations in human milk with 46% [13].

Vitamin C: Vitamin C and its isolate ascorbic acid are natural antioxidant components present in human milk. They have efficacious antioxidant function due to its ability to donate electrons and thus protecting important biomolecules [38]. Six studies showed that Holder pasteurization decreased vitamin C between 12% to 40% and ascorbic acid between 16% to 26% [6,21,28–31]. Alternative thermal pasteurization (100 °C, 5 min) reduced vitamin C and ascorbic acid more severely compared to Holder pasteurization, with a decrease of 29% for vitamin C and 41% for ascorbic acid [30]. Studies on the effect of high-pressure treatment on vitamin C and ascorbic acid report conflicting results. One study found that high-pressure treatment (193 MPa, −20 °C) led to a decrease of approximately 11% in ascorbic acid [28], while another study (400–600 MPa, 5 min, 12 °C) found that it did not alter ascorbic acid concentration in human milk [29]. In addition, vitamin C levels were not influenced by high-pressure treatment [28,29]. With respect to ultraviolet-C treatment, one study found that pasteurization with a radiation dose of 173–740 J/L reduced vitamin C levels by 15% to 35% [21].

Vitamin E: Vitamin E can be divided into four different tocopherols (α-, β-, γ-, and δ-tocopherol) and four different tocotrienols. Vitamin E has an important antioxidant function as it protects polyunsaturated fatty acids and other substances from peroxidation [20]. The most abundant tocopherol in human milk is α-tocopherol, which is the most biologically active form of vitamin E [36]. Several studies investigated the effect of pasteurization on tocopherols in human milk. Four studies investigated the effect of Holder pasteurization on α-tocopherol: two showed a reduction [20,30], while the other two showed no effect [29,31]. One study showed that heating at 100 °C for 5 min also reduced the α-tocopherol activity in human milk [30]. High-pressure treatment reduced α-tocopherol between 21–27% at 600 MPa for 3–6 min [20], while another study showed no effect of high-pressure treatment at 400, 500, and 600 MPa. A combination of thermal treatment at 65 and 80 °C at any pressure showed a reduction in α-tocopherol [34].

The studies investigating the effect of pasteurization on γ- and δ-tocopherol also showed conflicting results. Three studies investigated the effect of Holder pasteurization on γ-tocopherol, of which two showed a reduction between 13–47% [20,30] and one showed no effect [29]. Two studies investigated the effect of Holder pasteurization on δ-tocopherol [20], with one showing a reduction of 33% and the other no effect [29]. After alternative thermal pasteurization (100 °C for 5 min) a reduction was found in γ-tocopherol [30]. Two studies investigated the effect of high pressure on γ- and δ-tocopherol, of which one study showed a reduction only at a specific pressure and the other showed no effect [20,29]. A combination of thermal and high pressure treatment also reduced both γ- and δ-tocopherol in human milk [34].

Vitamin A: Vitamin A consists of provitamin A (carotenoids: α-carotene and β-carotene) and non-provitamin A (lutein, zeaxanthin, and lycopene). Provitamin A is the most important for vitamin A activity, however, non-provitamin A also has antioxidant properties [39]. Several studies investigated the effect of pasteurization on provitamin A.

No effect of Holder pasteurization on vitamin A or β-carotene specifically was found [14,31]. High-pressure treatment did not influence β-carotene activity, and microwave heating (35–40 °C, 15–30 s) did not influence α- or β-carotene [19,40]. Moreover, no differences in vitamin A stability were found after the thermosonication pasteurization of human milk samples [14].

Several studies investigated the effect of pasteurization on non-provitamin A. Holder pasteurization reduced lutein and zeaxanthin by 16%, while it increased lycopene by 9% [19]. High-pressure treatment (100–600 MPa) reduced the concentrations of lutein and zeaxanthin by 40% to 60%, while it increased the concentration of lycopene by 6% to 14% [19]. After microwave heating, no effects on the concentrations of lycopene and lutein were observed.

Other antioxidants: Some trace elements contribute to the antioxidant function of human milk. For example, zinc and copper are cofactors in the superoxide dismutase reaction [32]. Selenium also contributes to antioxidant properties as it is part of the antioxidant selenoprotein enzyme that protects the cell against free radicals [32]. Moreover, selenium is a component of glutathione peroxidase [32]. Holder pasteurization did not affect zinc, copper, or selenium levels in human milk [32].

Table 1. Results of different pasteurization methods on antioxidant properties in human milk: total antioxidant capacity.

Antioxidant Components	Pasteurized Milk Samples	Holder Pasteurization	Alternative Thermal Pasteurization	Ultraviolet-C Treatment	Thermosonication	References
ABTS	N = 8	↓ (30.5%)	-	-	-	[23]
	N = 21 *	=	-	-	-	[24]
	N = 20	=	-	55, 173, 355, 544 and 740 J/L): = (N = 18)	-	[21]
	N = 10	↓ (% n.r.)	-	-	↑ (% n.r.)	[14]
	N = 7	=	20 °C, 193 MPa: =	-	-	[25]
DPPH	N = 8	=	-	-	-	[23]
	N = 10	↓ (% n.r.)	-	-	↑ (% n.r.)	[14]
FRAP	N = 21 *	=	-	-	-	[24]
ORAC	N = 10	=	-	-	-	[6]
ORP	N = 10	=	-	-	-	[6]
Unknown	N = 31	↓ (67%)	75 °C, 15 s: =	-	-	[13]
	N = 3	-	−15, 0, 10, 30 and 50 °C combined with 200, 400, 600 and 800 MPa for 1 s: = −15 °C, 400 MPa and −30 °C, 800 MPa: ↑	-	-	[26]

Abbreviations: MPa = Megapascal; J/L = Joule per liter; ABTS = 2,2′-azino-bis (3-ethylbenzothiazoline-6-sulfonzuur); DPPH = 2,2-difenyl-1-picrylhydrazyl; FRAP = ferric reducing ability of plasma; ORAC = oxygen radical absorbance capacity; ORP = oxidation-reduction potential; n.r. = not reported. The N indicates the number of pasteurized human milk samples. The effect of pasteurization is displayed compared with untreated human milk samples. * In this study, pasteurized human milk samples were compared with human milk samples from a control group instead of within-person comparisons. The symbols in this table should be read as: ↑ increase, = no effect, ↓ decrease.

Table 2. Results of different pasteurization methods on antioxidant properties in human milk: enzymatic antioxidants.

Antioxidant Component	Pasteurized Milk Samples	Holder Pasteurization	Alternative Thermal Pasteurization	Microwave Heating	Ultraviolet-C Treatment	References
Superoxide dismutase	N = 10	↓ (% n.r.)	-	-	-	[6]
	N = 10	=	66 °C, 0–30 min: = 70 °C, 0–20 min: = 70 °C, 30 min: ↓ (35%)	70 °C 1 min: ↑ (34%)	-	[22]
	N = 7	-	<0 °C, 193 MPa: ↑ (57%)	-	-	[25]

Table 2. *Cont.*

Antioxidant Component	Pasteurized Milk Samples	Holder Pasteurization	Alternative Thermal Pasteurization	Microwave Heating	Ultraviolet-C Treatment	References
Catalase	N = 10	↓ (57%)	66 °C, 0–30 min: ↓ (59.7–81.9%) 70 °C, 0–30 min: ↓ (58.4–86.9%)	62.5 °C 1–10 min: ↓ (33–39%) 66 °C 1–10 min: ↓ (39–48%) 70 °C 1–10 min: ↓ (38–52%)	-	[22]
Glutathione Peroxidase	N = 8	↓ (60%)	-	-	85–740 J/L: =	[21]
	N = 17	↓ (63%)	75 °C, 15s: ↓ (62%)	-	-	[13]
	N = 10	↓ (% n.r.)	-	-	-	[6]
	N = 21	↓ (23–70%)	-	-	-	[27]
	N = 10	↓ (42%)	66 °C, 0–30 min: ↓ (41.1–45.8%) 70 °C, 0–30 min: ↓ (44.4–56.1%)	62.5 °C 1–10 min: ↓ (11–44%) 66 °C 1–10 min: ↓ (28–42%) 70 °C 1–10 min: ↓ (31–53%)	-	[22]

Abbreviations: MPa = Megapascal; J/L = Joule per liter; n.r. = not reported. The N indicates the number of pasteurized human milk samples. The effect of pasteurization is displayed compared with untreated human milk samples. The symbols in this table should be read as: ↑ increase, = no effect, ↓ decrease.

Table 3. Results of different pasteurization methods on antioxidant properties in human milk: non-enzymatic antioxidants.

Antioxidant Components	Pasteurized Milk Samples	Holder Pasteurization	Alternative Thermal Pasteurization	High Pressure Treatment	Microwave Heating	Ultraviolet-C Treatment	Thermosonication	References
Glutathione	N = 31	↓ (46%)	75 °C, 15 s: =	-	-	-	-	[13]
Vitamin C	N = 7	↓ (35%)	-	−20 °C, 193 MPA: =	-	-	-	[28]
	N = 10	↓ (19.9%)	-	400, 500 and 600 MPa: =	-	-	-	[29]
	N = 5	↓ (36%)	-	-	-	-	-	[31]
	N = 11	↓ (38.4%)	-	-	-	85 J/L: = 173–740 J/L: ↓ (15–35%)	-	[21]
	N = 10	↓ (12%)	100 °C, 5 min: ↓ (29%)	-	-	-	-	[30]
Ascorbic acid	N = 7	↓ (24%)	-	−20 °C, 193 MPA: ↓	-	-	-	[28]
	N = 10	↓ (16.2%)	-	400, 500 and 600 MPa: =	-	-	-	[29]
	N = 10	↓ (% n.r.)	-	-	-	-	-	[6]
	N = 10	↓ (26%)	100 °C, 5 min: ↓ (41%)				-	[30]
Vitamin E: α-Tocopherol	N = 3	↓ (25%)	-	400 MPa, 3/6 min: = 600 MPa, 3/6 min: ↓ (21%, 27%)	-	-	-	[20]
	N = 9	=	-	-	-	-	-	[31]
	N = 10	=	-	400, 500 and 600 MPa: =	-	-	-	[29]
	N = 10	↓ (13–17%)	100 °C, 5 min: ↓ (32–34%)	-	-	-	-	[30]
	N = 6	-	50 °C, 300 MPa, 600 MPa, 900 MPa: = 65/80 °C, any pressure: ↓	-	-	-	-	[34]
Vitamin E: γ-Tocopherol	N = 3	↓ (47%)	-	400, 600 MPa, 3/6 min: ↓ (26–47%)	-	-	-	[20]
	N = 10	=	-	400, 500 and 600 MPa: =	-	-	-	[29]
	N = 10	↓ (13–17%)	100 °C, 5 min: ↓ (32–34%)	-	-	-	-	[30]
	N = 6	-	50, 65 and 80 °C, 300/600 MPa: ↓ 50 °C, 900 MPa: = 65 and 85 °C, 900 MPa: ↓	-	-	-	-	[34]
Vitamin E: δ-Tocopherol	N = 3	↓ (33%)	-	400 MPa, 3–6 min: = 600 MPa, 3–6 min: ↓ (25, 33%)	-	-	-	[20]
	N = 10	=	-	at 400, 500 and 600 MPa: =	-	-	-	[29]
	N = 6	-	50, 65 and 80 °C, 300/600 MPa: ↓ 50 °C, 900 MPa: = 65 and 85 °C, 900 MPa: ↓	-	-	-	-	[34]
Vitamine A:	N = 9	=	-	-	-	-	-	[31]
	N = 10	=	-	-	-	-	=	[14]
Vitamine A: β-carotene	N = 18–24	=	-	=	-	-	-	[19]
	N = 30	-	-	-	35–40 °C, 15–30s: =	-	-	[40]
Vitamine A: α-Carotene	N = 30	-	-	-	35–40 °C, 15–30 s: =	-	-	[40]

Table 3. *Cont.*

Antioxidant Components	Pasteurized Milk Samples	Holder Pasteurization	Alternative Thermal Pasteurization	High Pressure Treatment	Microwave Heating	Ultraviolet-C Treatment	Thermosonication	References
Vitamine A: Lutein + Zeaxanthin	N = 18–24	↓ (15.8%)	-	600 MPa: ↓ (60.2%) 200 + 400 MPa: ↓ (47.1%) 200 + 600 MPa: ↓ (57.5%) 100 + 600 MPa: ↓ (40%) 450 MPa: ↓ (57.6%)	-	-	-	[19]
Vitamine A: Lycopene:	N = 30	-	-	-	35–40 °C, 15–30 s: =	-	-	[40]
Vitamine A: Luteine	N = 30	-	-	-	35–40 °C: 15–30 s: =	-	-	[40]
Trace elements	N = 16	Zinc: = Copper: = Selenium: =	-	-	-	-	-	[32]

Abbreviations: MPa = megapascal; J/L = joule per liter; n.r. = not reported. The N indicates the number of pasteurized human milk samples. The effect of pasteurization is displayed compared with untreated human milk samples. The symbols in this table should be read as: = no effect, ↓ decrease.

4. Discussion

This literature review summarized the current evidence on the influence of pasteurization on the antioxidant properties of human milk and compared different pasteurization methods. Overall, several studies have been conducted on investigating different human milk components and these studies suggest that Holder pasteurization reduces the antioxidant properties of human milk. Alternative pasteurization methods seem promising as less reduction is observed in several studies.

Currently, the most common pasteurization method is Holder pasteurization, in which milk is exposed to a temperature of approximately 62.5 °C for at least 30 min [41]. Most studies investigating the effect of Holder pasteurization on the total antioxidant capacity of human milk showed no effect, whereas others, including the study with the largest sample size (13), did show a reduction of up to 67% of the total antioxidant capacity. The differences between studies could partly be due to the different methods used to measure the total antioxidant capacity. When evaluating the separate antioxidant components that contribute to the total antioxidant capacity of human milk, in most studies, a reduction was observed after Holder pasteurization. In particular, catalase, glutathione peroxidase, glutathione, vitamin C, and ascorbic acid decreased after Holder pasteurization in all of those studies. Studies on the effect of Holder pasteurization on the other antioxidant components of human milk were controversial in their results.

An explanation for the reduction of antioxidant activity in Holder-pasteurized donor milk might be the denaturation of proteins during the heating process [42]. Holder pasteurization is recommended by all international human milk bank guidelines. However, as this process seems to reduce some of the beneficial effects of human milk, safe alternatives should be considered. To date, studies on the effects of other pasteurization techniques on human milk components and, specifically, antioxidant function, are scarce. The limited results demonstrate that alternative pasteurization methods including high-pressure treatment, microwave heating, and ultraviolet-C treatment seem promising and cause less of a reduction in the antioxidant components, especially enzymatic agents. These results are in line with the fact that enzymes denature during the heating process. Moreover, non-enzymatic agents such as vitamins were also less affected by alternative pasteurization methods, which might be due to the size of the molecules, as small molecules are less affected by, for example, high pressure [29]. Remarkably, some studies even show an increase in the antioxidant components of human milk after alternative pasteurization methods. For example, after high-pressure treatment and microwave heating an increase in superoxide dismutase was found. A possible explanation might be that neutrophils in human milk are destroyed under high pressure, causing a release of superoxide dismutase from these cells [34]. During microwave heating, the energy transfer between the electromagnetic field and protein domains might change the enzymatic properties and subsequently increase the reactivity of enzymes, for example, superoxide dismutase [43]. Altogether, this confirms the main concerns about the Holder pasteurization of donor milk and the need for future strategies for which the above-described pasteurization methods seem promising.

As preterm infants are especially susceptible to damage from oxidative stress, the antioxidant properties of human milk are extremely important for this specific group. These infants receive donor milk relatively frequently in the first days of life, when the amount of oxidative stress is at its highest. Therefore, knowledge on the effect of pasteurization methods on the antioxidant activity is important. Moreover, it has been demonstrated that preterm infants receiving pasteurized human milk have poorer growth and developmental outcomes compared to infants receiving unpasteurized human milk [18,44]. This is most likely due to the effect of pasteurization on multiple aspects of human milk and not only on the effect on antioxidants. Thus, it is important to investigate the effect of pasteurization on other human milk components as well, for instance the effect of pasteurization on nutrients, immunological components and the human milk microbiome.

To draw conclusions and enable unbiased, quantitative comparisons of the effect of pasteurization methods on the antioxidant properties of human milk to guide and advise human donor milk banks, future studies should take the following aspects into account. First, study protocols and procedures should be standardized (e.g., sample origin, storage conditions, methods for measuring the TAC). Second, studies should compare different pasteurization techniques within the same human milk samples as antioxidant capacity of human milk is influenced by several other factors, for example, maternal diet [45]. Third, it appears that some antioxidants were investigated more extensively than others. For instance, some short chain fatty acids and amino acids are, next to their nutritional value, also known for their antioxidant properties and were not described in this review [46,47]. Fourth, investigating the functionality of antioxidant components in human milk should be preferred over measuring the concentration. Moreover, studies investigating oxidative stress-related outcomes between preterm infants receiving unpasteurized milk and donor milk treated with different pasteurization methods are currently lacking. Randomized controlled trials should be setup to investigate the effect of pasteurization on these outcomes to draw conclusions on the clinical relevance and to optimize current human milk bank pasteurization guidelines.

5. Conclusions

As many human milk components remain relatively unaffected by pasteurization and cow's milk-based formula does not contain many bioactive components, donor milk is still considered the best alternative for a mother's milk. Research on the effects of pasteurization method on the antioxidant properties of human milk is scarce. In general, Holder pasteurization seems to reduce the antioxidant properties of human milk, specifically, the activity of certain antioxidant molecules. Whether this reduction is clinically relevant remains to be determined. Alternative pasteurization methods seem promising as less reduction in the TAC was observed in several studies. More research is necessary to improve knowledge on the effect of different pasteurization methods on human milk antioxidants to guide human milk banks in their pasteurization processes to optimize early-life nutrition and, thereby also, the health outcomes of preterm infants.

Author Contributions: Conceptualization, H.G.J., E.J.M.R., J.B.v.G. and B.J.v.K.; methodology and literature search, H.G.J., E.J.M.R., G.L.B. and B.J.v.K., writing—original draft preparation, H.G.J., E.J.M.R. and B.J.v.K.; writing—review and editing, H.G.J., E.J.M.R., G.L.B., C.H.P.v.d.A., A.K., J.B.v.G. and B.J.v.K. All authors have read and agreed to the published version of the manuscript.

Funding: This research received no external funding.

Conflicts of Interest: J.B.v.G. is the founder and director of the Dutch National Human Milk Bank and a member of the National Health Council. J.B.v.G. has been a member of the National Breastfeeding Council from March 2010 to March 2020.

Appendix A

Table A1. Search strategy in PubMed.

Search	Query	Results
#4	#1 AND #2 AND #3	146
#3	"Pasteurization" [Mesh] OR "pasteuriz*" [tiab] OR "pasteuris*" [tiab] OR "Ultrapasteuriz*" [tiab] OR "heat treat*" [tiab] OR "high pressure" [tiab] OR "radiation treat*" [tiab] OR "thermal treat*" [tiab] OR "proces*" [tiab]	2,435,342
#2	"Milk, Human" [Mesh] OR "Breast Feeding" [Mesh] OR "Bottle Feeding" [Mesh] OR "Colostrum" [Mesh] OR "breast milk*" [tiab] OR "human milk*" [tiab] OR "breastmilk*" [tiab] OR "Breastfeeding*" [tiab] OR "Breast feeding*" [tiab] OR "Breast Fed" [tiab] OR "milk shar*" [tiab] OR "Wet Nursing" [tiab] OR "Bottlefeeding*" [tiab] OR "Bottlefed" [tiab] OR "donor milk" [tiab] OR "mothers own milk" [tiab] OR "colostrum" [tiab]	85,903
#1	"Antioxidants" [Mesh] OR "Antioxidants" [Pharmacological Action] OR "Oxidation-Reduction" [Mesh] OR "Oxidative Stress" [Mesh] OR "antioxida*" [tiab] OR "anti oxida*" [tiab] OR "Redox" [tiab] OR "reduction oxida*" [tiab] OR "oxidative stress" [tiab]	974,155

Table A2. Search strategy in Embase.com (accessed on 27 May 2021).

Search	Query	Results
#5	#4 NOT ('conference abstract'/it OR 'conference paper'/it)	134
#4	#1 AND #2 AND #3	172
#3	'pasteurization'/exp OR ('pasteuriz*' OR 'pasteuris*' OR 'Ultrapasteuriz*' OR 'heat treat*' OR 'high pressure' OR 'radiation treat*' OR 'thermal treat*' OR proces*'):ti,ab,kw	1453
#2	'breast milk'/exp OR 'breast feeding'/exp OR 'bottle feeding'/exp OR 'colostrum'/exp OR ('breast milk*' OR 'human milk*' OR 'breastmilk*' OR 'Breastfeeding*' OR 'Breast feeding*' OR 'Breast Fed' OR 'milk shar*' OR 'Wet Nursing' OR 'Bottlefeeding*' OR 'Bottlefed' OR 'donor milk' OR 'mothers own milk' OR 'colostrum'):ti,ab,kw	107,230
#1	'antioxidant'/exp OR 'oxidation reduction reaction'/exp OR 'oxidative stress'/exp OR ('antioxida*' OR 'anti oxida*' OR 'Redox' OR 'reduction oxida*' OR 'oxidative stress'):ti,ab,kw	805,784

Table A3. Search strategy in Cumulative Index to Nursing and Allied Health Literature (CINAHL).

Search	Query	Results
#4	#1 AND #2 AND #3	17
#3	(MH "Pasteurization") OR TI("pasteuriz*" OR "pasteuris*" OR "Ultrapasteuriz*" OR "heat treat*" OR "high pressure" OR "radiation treat*" OR "thermal treat*" OR "proces*") OR AB("pasteuriz*" OR "pasteuris*" OR "Ultrapasteuriz*" OR "heat treat*" OR "high pressure" OR "radiation treat*" OR "thermal treat*" OR "proces*") OR KW("pasteuriz*" OR "pasteuris*" OR "Ultrapasteuriz*" OR "heat treat*" OR "high pressure" OR "radiation treat*" OR "thermal treat*" OR "proces*")	339,674
#2	(MH "Milk, Human+") OR (MH "Breast Feeding+") OR (MH "Bottle Feeding") OR (MH "Colostrum") OR TI("breast milk*" OR "human milk*" OR "breastmilk*" OR "Breastfeeding*" OR "Breast feeding*" OR "Breast Fed" OR "milk shar*" OR "Wet Nursing" OR "Bottlefeeding*" OR "Bottlefed" OR "donor milk" OR "mothers own milk" OR "colostrum") OR AB("breast milk*" OR "human milk*" OR "breastmilk*" OR "Breastfeeding*" OR "Breast feeding*" OR "Breast Fed" OR "milk shar*" OR "Wet Nursing" OR "Bottlefeeding*" OR "Bottlefed" OR "donor milk" OR "mothers own milk" OR "colostrum") OR KW("breast milk*" OR "human milk*" OR "breastmilk*" OR "Breastfeeding*" OR "Breast feeding*" OR "Breast Fed" OR "milk shar*" OR "Wet Nursing" OR "Bottlefeeding*" OR "Bottlefed" OR "donor milk" OR "mothers own milk" OR "colostrum")	36,872
#1	(MH "Antioxidants+") OR (MH "Oxidation-Reduction+") OR (MH "Oxidative Stress") OR TI("antioxida*" OR "anti oxida*" OR "Redox" OR "reduction oxida*" OR "oxidative stress") OR AB("antioxida*" OR "anti oxida*" OR "Redox" OR "reduction oxida*" OR "oxidative stress") OR KW("antioxida*" OR "anti oxida*" OR "Redox" OR "reduction oxida*" OR "oxidative stress")	56,682

Table A4. Search strategy in Clarivate Analytics/Web of Science Core Collection.

Search	Query	Results
#4	#1 AND #2 AND #3	166
#3	TS = ("pasteuriz*" OR "pasteuris*" OR "Ultrapasteuriz*" OR "heat treat*" OR "high pressure" OR "radiation treat*" OR "thermal treat*" OR "proces*")	5,760,107
#2	TS = ("breast milk*" OR "human milk*" OR "breastmilk*" OR "Breastfeeding*" OR "Breast feeding*" OR "Breast Fed" OR "milk shar*" OR "Wet Nursing" OR "Bottlefeeding*" OR "Bottlefed" OR "donor milk" OR "mothers own milk" OR "colostrum")	71,705
#1	TS = ("antioxida*" OR "anti oxida*" OR "Redox" OR "reduction oxida*" OR "oxidative stress")	838,494

Table A5. Search strategy in Wiley/Cochrane Library.

Search	Query	Results
#3	#1 AND #2	4
#2	("breast NEXT milk*" OR "human NEXT milk*" OR "breastmilk*" OR "Breastfeeding*" OR "Breast NEXT feeding*" OR "Breast NEXT Fed" OR "milk NEXT shar*" OR "Wet NEXT Nursing" OR "Bottlefeeding*" OR "Bottlefed" OR "donor NEXT milk" OR "mothers NEXT own NEXT milk" OR "colostrum"):ti,ab,kw	6481
#1	("antioxida*" OR "anti NEXT oxida*" OR "Redox" OR "reduction NEXT oxida*" OR "oxidative NEXT stress"):ti,ab,kw	759

References

1. Moore, T.A.; Ahmad, I.M.; Zimmerman, M.C. Oxidative stress and preterm birth: An integrative review. *Biol. Res. Nurs.* **2018**, *20*, 497–512. [CrossRef] [PubMed]
2. Vento, M.; Cheung, P.Y.; Aguar, M. The first golden minutes of the extremely-low-gestational-age neonate: A gentle approach. *Neonatology* **2009**, *95*, 286–298. [CrossRef] [PubMed]
3. Ahola, T.; Fellman, V.; Kjellmer, I.; Raivio, K.O.; Lapatto, R. Plasma 8-isoprostane is increased in preterm infants who develop bronchopulmonary dysplasia or periventricular leukomalacia. *Pediatr. Res.* **2004**, *56*, 88–93. [CrossRef] [PubMed]
4. Saugstad, O.D. Oxygen and retinopathy of prematurity. *J. Perinatol.* **2006**, *26*, S46–S50. [CrossRef]
5. Buonocore, G.; Perrone, S.; Bracci, R. Free radicals and brain damage in the newborn. *Biol. Neonate* **2001**, *79*, 180–186. [CrossRef] [PubMed]
6. Marinković, V.; Ranković-Janevski, M.; Spasić, S.; Nikolić-Kokić, A.; Lugonja, N.; Djurović, D.; Miletić, S.; Vrvić, M.M.; Spasojević, I. Antioxidative activity of colostrum and human milk: Effects of pasteurization and storage. *J. Pediatr. Gastroenterol. Nutr.* **2016**, *62*, 901–906. [CrossRef] [PubMed]
7. Szlagatys-Sidorkiewicz, A.; Zagierski, M.; Jankowska, A.; Łuczak, G.; Macur, K.; Bączek, T.; Korzon, M.; Krzykowski, G.; Martysiak-Żurowska, D.; Kamińska, B. Longitudinal study of vitamins A, E and lipid oxidative damage in human milk throughout lactation. *Early Hum. Dev.* **2012**, *88*, 421–424. [CrossRef]
8. Friel, J.K.; Martin, S.M.; Langdon, M.; Herzberg, G.R.; Buettner, G.R. Milk from mothers of both premature and full-term infants provides better antioxidant protection than does infant formula. *Pediatr. Res.* **2002**, *51*, 612–618. [CrossRef] [PubMed]
9. Lugonja, N.; Spasić, S.D.; Laugier, O.; Nikolić-Kokić, A.; Spasojević, I.; Oreščanin-Dušić, Z.; Vrvić, M.M. Differences in direct pharmacologic effects and antioxidative properties of mature breast milk and infant formulas. *Nutrition* **2013**, *29*, 431–435. [CrossRef]
10. Bjorklund, G.; Chirumbolo, S. Role of oxidative stress and antioxidants in daily nutrition and human health. *Nutrition* **2017**, *33*, 311–321. [CrossRef] [PubMed]
11. Johannes ESPGHAN Committee on Nutrition; Arslanoglu, S.; Corpeleijn, W.; Moro, G.; Braegger, C.; Campoy, C.; Colomb, V.; Decsi, T.; Domellof, M.; Fewtrell, M.; et al. Donor human milk for preterm infants: Current evidence and research directions. *J. Pediatr. Gastroenterol. Nutr.* **2013**, *57*, 535–542. [CrossRef]
12. Tully, D.B.; Jones, F.; Tully, M.R. Donor milk: What's in it and what's not. *J. Hum. Lact.* **2001**, *17*, 152–155. [CrossRef]
13. Silvestre, D.; Miranda, M.; Muriach, M.; Almansa, I.; Jareño, E.; Romero, F.J. Antioxidant capacity of human milk: Effect of thermal conditions for the pasteurization. *Acta Paediatr.* **2008**, *97*, 1070–1074. [CrossRef] [PubMed]
14. Parreiras, P.M.; Vieira Nogueira, J.A.; Rodrigues da Cunha, L.; Passos, M.C.; Gomes, N.R.; Breguez, G.S.; Falco, T.S.; Bearzoti, E.; Menezes, C.C. Effect of thermosonication on microorganisms, the antioxidant activity and the retinol level of human milk. *Food Control* **2020**, *113*, 107172. [CrossRef]
15. Moro, G.E.; Billeaud, C.; Rachel, B.; Calvo, J.; Cavallarin, L.; Christen, L.; Escuder-Vieco, D.; Gaya, A.; Lembo, D.; Wesolowska, A.; et al. Processing of donor human milk: Update and recommendations from the european milk bank association (EMBA). *Front. Pediatr.* **2019**, *7*, 49. [CrossRef] [PubMed]

16. Italian Association of Human Milk Banks Associazione Italiana Banche del Latte Umano Donato; Arslanoglu, S.; Bertino, E.; Tonetto, P.; De Nisi, G.; Ambruzzi, A.M.; Biasini, A.; Profeti, C.; Spreghini, M.R.; Moro, G.E. Guidelines for the establishment and operation of a donor human milk bank. *J. Matern. Fetal Neonatal Med.* **2010**, *23*, 1–20. [CrossRef] [PubMed]

17. Peila, C.; Moro, G.E.; Bertino, E.; Cavallarin, L.; Giribaldi, M.; Giuliani, F.; Cresi, F.; Coscia, A. The effect of holder pasteurization on nutrients and biologically-active components in donor human milk: A review. *Nutrients* **2016**, *8*, 477. [CrossRef]

18. Montjaux-Regis, N.; Cristini, C.; Arnaud, C.; Glorieux, I.; Vanpee, M.; Casper, C. Improved growth of preterm infants receiving mother's own raw milk compared with pasteurized donor milk. *Acta Paediatr.* **2011**, *100*, 1548–1554. [CrossRef]

19. Wesolowska, A.; Brys, J.; Barbarska, O.; Strom, K.; Szymanska-Majchrzak, J.; Karzel, K.; Pawlikowska, E.; Zielinska, M.A.; Hamulka, J.; Oledzka, G. Lipid Profile, Lipase bioactivity, and lipophilic antioxidant content in high pressure processed donor human milk. *Nutrients* **2019**, *11*, 1972. [CrossRef]

20. Delgado, F.; Cava, R.; Delgado, J.; Ramírez, R. Tocopherols, fatty acids and cytokines content of Holder pasteurised and high-pressure processed human milk. *Dairy Sci. Technol.* **2014**, *94*, 145–156. [CrossRef]

21. Martysiak-Żurowska, D.; Puta, M.; Kotarska, J.; Cybula, K.; Malinowska-Pańczyk, E.; Kołodziejska, I. The effect of UV-C irradiation on lipids and selected biologically active compounds in human milk. *Int. Dairy J.* **2017**, *66*, 42–48. [CrossRef]

22. Martysiak-Żurowska, D.; Puta, M.; Kiełbratowska, B. The effect of convective heating and microwave heating on antioxidant enzymes in pooled mature human milk. *Int. Dairy J.* **2019**, *91*, 41–47. [CrossRef]

23. Nogueira, J.A.V.; Santos, M.M.D.O.; Passos, M.C.; Cunha, L.R.D.; Parreiras, P.M.; Menezes, C.C. Stability of Antioxidant Capacity of Human Milk after Freezing and Pasteurization. *J. Food Nutr. Res.* **2018**, *6*, 450–455. [CrossRef]

24. Lorençoni, M.F.; Silva, R.S.; Azevedo Júnior, R.; Fronza, M. Effect of pasteurization on the antioxidant and oxidant properties of human milk. *Rev. Paul. Pediatr.* **2021**, *39*, e2019165. [CrossRef]

25. Malinowska-Pańczyk, E.; Martysiak-Żurowska, D.; Puta, M.; Kusznierewicz, B. The effect of high pressure and subzero temperature on the microflora and selected components of human milk. In Proceedings of the 55th EHPRG Meeting: High Pressure Science and Technology, Poznań, Poland, 3–8 September 2017.

26. Ramírez, R.; Garrido, M.; Rocha-Pimienta, J.; García-Parra, J.; Delgado-Adámez, J. Immunological components and antioxidant activity in human milk processed by different high pressure-thermal treatments at low initial temperature and flash holding times. *Food Chem.* **2021**, *343*, 128546. [CrossRef] [PubMed]

27. Guerra, A.F.; Mellinger-Silva, C.; Rosenthal, A.; Luchese, R.H. Hot topic: Holder pasteurization of human milk affects some bioactive proteins. *J. Dairy Sci.* **2018**, *101*, 2814–2818. [CrossRef] [PubMed]

28. Martysiak-Żurowska, D. Effect of high pressure and sub-zero temperature on total antioxidant capacity and the content of vitamin c, fatty acids and secondary products of lipid oxidation in human milk. *Pol. J. Food Nutr. Sci.* **2017**, *67*, 117–122. [CrossRef]

29. Moltó-Puigmartí, C.; Permanyer, M.; Castellote, A.I.; López-Sabater, M.C. Effects of pasteurisation and high-pressure processing on vitamin C, tocopherols and fatty acids in mature human milk. *Food Chem.* **2011**, *124*, 697–702. [CrossRef]

30. Romeu-Nadal, M.; Castellote, A.; Gayá, A.; López-Sabater, M. Effect of pasteurisation on ascorbic acid, dehydroascorbic acid, tocopherols and fatty acids in pooled mature human milk. *Food Chem.* **2008**, *107*, 434–438. [CrossRef]

31. Van Zoeren-Grobben, D.; Schrijver, J.; Van den Berg, H.; Berger, H.M. Human milk vitamin content after pasteurisation, storage, or tube feeding. *Arch. Dis. Child.* **1987**, *62*, 161–165. [CrossRef]

32. Mohd-Taufek, N.; Cartwright, D.; Davies, M.; Hewavitharana, A.K.; Koorts, P.; McConachy, H.; Shaw, P.N.; Sumner, R.; Whitfield, K. The effect of pasteurization on trace elements in donor breast milk. *J. Perinatol.* **2016**, *36*, 897–900. [CrossRef]

33. Wesolowska, A.; Sinkiewicz-Darol, E.; Barbarska, O.; Bernatowicz-Lojko, U.; Borszewska-Kornacka, M.K.; van Goudoever, J.B. Innovative techniques of processing human milk to preserve key components. *Nutrients* **2019**, *11*, 1169. [CrossRef] [PubMed]

34. Delgado, F.J.; Contador, R.; Álvarez-Barrientos, A.; Cava, R.; Delgado-Adámez, J.; Ramírez, R. Effect of high pressure thermal processing on some essential nutrients and immunological components present in breast milk. *Innov. Food Sci. Emerg. Technol.* **2013**, *19*, 50–56. [CrossRef]

35. Rubio, C.P.; Hernández-Ruiz, J.; Martinez-Subiela, S.; Tvarijonaviciute, A.; Ceron, J.J. Spectrophotometric assays for total antioxidant capacity (TAC) in dog serum: An update. *BMC Vet. Res.* **2016**, *12*, 166. [CrossRef] [PubMed]

36. Živković, J.; Sunarić, S.; Trutić, N.; Denić, M.; Kocić, G.; Jovanović, T. Antioxidants and antioxidant capacity of human milk/Antioksidansi i antioksidativni kapacitet humanog mleka. *Acta Fac. Med. Naissensis* **2015**, *32*, 115–125. [CrossRef]

37. Ankrah, N.-A.; Appiah-Opong, R.; Dzokoto, C. Human breastmilk storage and the glutathione content. *J. Trop. Pediatr.* **2000**, *46*, 111–113. [CrossRef]

38. Carr, A.; Frei, B. Does vitamin C act as a pro-oxidant under physiological conditions? *FASEB J.* **1999**, *13*, 1007–1024. [CrossRef]

39. Levêques, A.; Oberson, J.M.; Tissot, E.A.; Redeuil, K.; Thakkar, S.K.; Campos-Giménez, E. Quantification of vitamins A, E, and K and carotenoids in submilliliter Volumes of human milk. *J. AOAC Int.* **2019**, *102*, 1059–1068. [CrossRef] [PubMed]

40. Tacken, K.J.; Vogelsang, A.; van Lingen, R.A.; Slootstra, J.; Dikkeschei, B.D.; van Zoeren-Grobben, D. Loss of triglycerides and carotenoids in human milk after processing. *Arch. Dis. Child.* **2009**, *94*, F447–F450. [CrossRef]

41. O'Connor, D.L.; Ewaschuk, J.B.; Unger, S. Human milk pasteurization: Benefits and risks. *Curr. Opin. Clin. Nutr. Metab. Care* **2015**, *18*, 269–275. [CrossRef]

42. Bischof, J.C.; He, X. Thermal stability of proteins. *Ann. N. Y. Acad. Sci.* **2005**, *1066*, 12–33. [CrossRef] [PubMed]

43. Mazumder, S.; Laskar, D.D.; Prajapati, D.; Roy, M.K. Microwave-induced enzyme-catalyzed chemoselective reduction of organic azides. *Chem. Biodivers.* **2004**, *1*, 925–929. [CrossRef] [PubMed]

44. Hard, A.L.; Nilsson, A.K.; Lund, A.M.; Hansen-Pupp, I.; Smith, L.E.H.; Hellstrom, A. Review shows that donor milk does not promote the growth and development of preterm infants as well as maternal milk. *Acta Paediatr.* **2019**, *108*, 998–1007. [CrossRef] [PubMed]
45. Oveisi, M.R.; Sadeghi, N.; Jannat, B.; Hajimahmoodi, M.; Behfar, A.O.; Jannat, F.; Mokhtarinasab, F. Human breast milk provides better antioxidant capacity than infant formula. *Iran. J. Pharm. Res.* **2010**, *9*, 445–449.
46. Elias, R.J.; Kellerby, S.S.; Decker, E.A. Antioxidant activity of proteins and peptides. *Crit. Rev. Food Sci. Nutr.* **2008**, *48*, 430–441. [CrossRef] [PubMed]
47. Yeap, S.K.; Beh, B.K.; Ali, N.M.; Yusof, H.M.; Ho, W.Y.; Koh, S.P.; Alitheen, N.B.; Long, K. Antistress and antioxidant effects of virgin coconut oil in vivo. *Exp. Ther. Med.* **2015**, *9*, 39–42. [CrossRef] [PubMed]

antioxidants

Review

Metabolomics to Diagnose Oxidative Stress in Perinatal Asphyxia: Towards a Non-Invasive Approach

Anne Lee Solevåg [1],*, Svetlana N. Zykova [2], Per Medbøe Thorsby [2] and Georg M. Schmölzer [3,4]

[1]   The Department of Paediatric and Adolescent Medicine, Oslo University Hospital, 0424 Nydalen, Norway
[2]   Biochemical Endocrinology and Metabolism Research Group, The Hormone Laboratory, Department of Medical Biochemistry, Oslo University Hospital, 0424 Nydalen, Norway; svezyk@ous-hf.no (S.N.Z.); pertho@ous-hf.no (P.M.T.)
[3]   Centre for the Studies of Asphyxia and Resuscitation, Neonatal Research Unit, Royal Alexandra Hospital, Edmonton, AB 23821, Canada; georg.schmoelzer@me.com
[4]   Department of Pediatrics, Faculty of Medicine and Dentistry, University of Alberta, Edmonton, AB 23821, Canada
*   Correspondence: a.l.solevag@medisin.uio.no; Tel.: +47-4146-9314

Abstract: There is a need for feasible and non-invasive diagnostics in perinatal asphyxia. Metabolomics is the study of small molecular weight products of cellular metabolism that may, directly and indirectly, reflect the level of oxidative stress. Saliva analysis is a novel approach that has a yet unexplored potential in metabolomics in perinatal asphyxia. The aim of this review was to give an overview of metabolomics studies of oxidative stress in perinatal asphyxia, particularly searching for studies analyzing non-invasively collected biofluids including saliva. We searched the databases PubMed/Medline and included 11 original human and 4 animal studies. In perinatal asphyxia, whole blood, plasma, and urine are the most frequently used biofluids used for metabolomics analyses. Although changes in oxidative stress-related salivary metabolites have been reported in adults, the utility of this approach in perinatal asphyxia has not yet been explored. Human and animal studies indicate that, in addition to antioxidant enzymes, succinate and hypoxanthine, as well acylcarnitines may have discriminatory diagnostic and prognostic properties in perinatal asphyxia. Researchers may utilize the accumulating evidence of discriminatory metabolic patterns in perinatal asphyxia to develop bedside methods to measure oxidative stress metabolites in perinatal asphyxia. Although only supported by indirect evidence, saliva might be a candidate biofluid for such point-of-care diagnostics.

Keywords: asphyxia neonatorum; non-invasive diagnostics; metabolomics; oxidative stress; saliva

Citation: Solevåg, A.L.; Zykova, S.N.; Thorsby, P.M.; Schmölzer, G.M. Metabolomics to Diagnose Oxidative Stress in Perinatal Asphyxia: Towards a Non-Invasive Approach. *Antioxidants* **2021**, *10*, 1753. https://doi.org/10.3390/antiox10111753

Academic Editors: Julia Kuligowski and Máximo Vento

Received: 1 October 2021
Accepted: 29 October 2021
Published: 2 November 2021

Publisher's Note: MDPI stays neutral with regard to jurisdictional claims in published maps and institutional affiliations.

1. Introduction

Failed placental gas exchange or deficient gas exchange in the lungs after birth may cause perinatal asphyxia with hypoxia and hypercapnia resulting in mixed metabolic and respiratory acidosis. Asphyxiated infants can present with severe cardiorespiratory compromise at birth and a need for cardiopulmonary resuscitation with supplementary oxygen. However, mild symptoms of asphyxia may also result in later morbidity and mortality in affected infants [1,2].

In perinatal asphyxia, balancing the harmful effects of iatrogenic hyperoxia ('oxidative stress') vs. anaerobic metabolism (continuing/prolonged hypoxia) is a complex task. Although humans have physiological and biochemical defense mechanisms to prevent hypoxia [3], defense mechanisms against hyperoxia are less developed in newborn infants, with the resulting oxidative stress potentially causing organ injury.

During asphyxia and resuscitation, disrupted cellular homeostasis causes significant metabolic changes [4], and studies of the metabolome may provide a pathophysiological 'snapshot' of the condition. Metabolomics is the study of small molecular weight (<1500 Da)

endogenous metabolites present in tissues or biofluids typically at concentrations above 1 uM, and may be used to directly and indirectly measure oxidative stress [5]. Measuring oxidative stress in the neonate could aid decisions to initiate time-critical interventions including therapeutic hypothermia but poses challenges due to difficult sampling as well as a need for quick and repeated analyses [6].

Saliva as a Promising Simple and Non-Invasively Collected Biofluid

In the search for therapeutic target molecules, saliva has been used as a matrix in genomics [7], proteomics [8], and metabolomics [9]. Non-invasive and safe real-time sampling, and ease of collection, handling, storage, and post-storage contribute to its attractiveness and relatively low costs. However, different collection methods might yield different analysis results. Thus, it is important to standardize the collection, e.g., either by passive drainage or by specific collection devices [8]. In newborn infants, saliva may have particular advantages over invasively collected biofluids including blood. Compared to blood sampling, saliva sampling is not painful and does not deplete patients with blood cells and nutrients.

The aim of this review paper is to provide an overview of metabolomics studies of oxidative stress in perinatal asphyxia, with particular emphasis on analysis methods and biofluids used. Measuring the level of oxidative stress might aid in diagnosing infants with significant perinatal asphyxia but a mild initial clinical presentation. We were particularly interested in evidence about feasible and non-invasive diagnostics, with saliva being a novel biofluid with yet unknown potential in perinatal asphyxia.

2. Materials and Methods

Search Strategy and Selection Criteria

We did a non-systematic search in PubMed/Medline with no limits on publication date. The search was finished in September 2021 and included the terms "metabolomics" AND "oxidative stress" AND "neonatal asphyxia" OR "perinatal asphyxia" OR "HIE" OR "hypoxic ischemic encephalopathy". Reference lists of relevant articles were hand searched for additional publications of interest. Only English-language, peer-reviewed studies were included. We excluded studies that analyzed oxidative stress in tissues, metabolomics studies with a lack of focus on oxidative stress, as well as review articles, case reports, and conference abstracts. We included studies in term and preterm infants, as well as animal studies. The search identified 11 original human studies and 4 original animal studies.

3. Results

3.1. Methods for Metabolomics Analyses

Metabolomics analyses can be targeted and untargeted [10], the latter investigating both known and unknown metabolites. Analytical methods for determination of small molecular weight metabolites in biofluids include (i) immunological methods including radioimmunoassay and enzyme-linked immunosorbent assay (targeted), (ii) high-resolution ^{1}H (proton) and ^{13}C nuclear magnetic resonance (NMR) spectroscopy [11] (targeted and untargeted), (iii) gas chromatography (GC) coupled to mass spectrometry (MS) (targeted and untargeted), (iv) (ultra-performance) liquid chromatography coupled to MS (LC-MS) (targeted and untargeted), (v) infrared and Raman spectroscopy [12,13], and (vi) capillary electrophoresis–time-of-flight mass spectrometry (CE–TOFMS) (semi-targeted).

MS-based assays are considered more accurate than immunological methods (Table 1), with LC-MS having advantages over GC-MS as it does not require as long derivatization processes that may cause measurement error. Moreover, ultra-high performance (UP) liquid chromatography tandem mass spectrometry (UPLC-MS/MS) enables rapid chromatographic separations. UPLC-MS/MS is characterized by good selectivity and sensitivity, and a high sample throughput [6]. The metabolomics MS-workflow for biofluids including saliva is presented in Figure 1.

Table 1. Advantages and disadvantages of mass spectrometry-based assays and immunoassays [14].

Method	Advantages	Disadvantages
Mass spectrometry	High selectivity	Limited to some laboratories
	High sensitivity	Limited sample throughput
	High throughput	Limited user friendliness, requires specialized personnel
	Requires low sample volumes	High equipment costs
	Multiplexing possible	Sample complexity issues
	Relatively low running costs	
	Not restricted to biomolecules	
	High intra- and inter-assay reproducibility	
Immunoassay	Low training requirements	Limited selectivity
	Kits available from commercial vendors	Limited analyte detection abilities
	Validated and approved	Requires relatively high sample volumes
	Relatively low-cost equipment	Relatively expensive reagents
	Relatively high throughput	Relatively high intra- and interassay and laboratory variability
	Relatively high sensitivity	Relatively long assay time

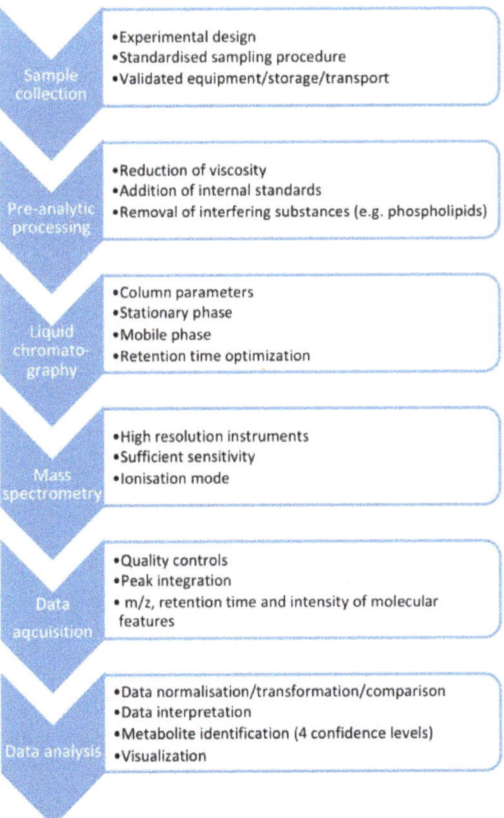

Figure 1. The metabolomics mass spectrometry workflow for biofluids including saliva.

3.2. Different Biofluids

3.2.1. Blood

Blood is the biological material for most diagnostic tests in clinical routine. Blood reflects the dynamic, real-time metabolic response to asphyxia and reoxygenation, and allows for quick analysis. However, blood sampling in neonates may be technically challenging, is invasive, and painful. Blood sampling may contribute to anemia in more premature infants as their blood volume is lower. Indeed, Sachse et al. [11] were able to perform ^{1}H NMR metabolomics analysis with only 250 µL of piglet plasma, and Sanchez-Illana et al. [6] used 100 µL plasma for UPLC-MS/MS analysis. Dekker et al. [15] used 300 µL full blood from premature infants.

An alternative blood sample option is umbilical cord blood, which is non-invasive and painless and may allow the collection of larger volumes to perform several analyses. However, umbilical cord blood only provides information until the time of birth. Thus, combined measurements of metabolites in umbilical cord blood and subsequently in the infant's blood are needed.

There is increasing evidence that preanalytical factors may heavily influence the results in metabolomics, both in the individual investigated and handling of the sample before metabolomic analyses. Currently, the need for access to advanced sample preparation is needed when conducting metabolomic studies [16]. Dried blood spot samples have been suggested for metabolomics analysis as they do not require preanalytical preparation, special storage, or freezing [17]. Several studies indicate the utility of blood spots for metabolomics analysis, with dried blood spots being largely equivalent to protein-precipitated plasma [18,19].

3.2.2. Urine

Although urine collection might be an easy alternative, the urine output in the first days after birth is typically low, and infants with perinatal asphyxia often have impaired renal function with oliguria or anuria. Urine for metabolomics analysis must be collected under sterile conditions because bacterial metabolism influences the urine metabolome, and samples must be frozen at −80 °C immediately after collection [20]. Urine is suitable for repeat analysis and assessment of markers of lipid peroxidation up to days or weeks after an insult or intervention, with the associated disadvantage of limited time resolution. Thus, urine analysis might not be the preferred option when the purpose is to institute time-critical interventions.

Sachse et al. [11] used 550 µL of piglet urine for ^{1}H NMR metabolomics analysis. They demonstrated a different time profile of plasma versus urine metabolites and concluded that renal handling of different metabolites may vary and needs to be considered if urine is used instead of plasma in metabolomic analyses.

3.2.3. Cerebrospinal Fluid

Cerebrospinal fluid (CSF) is used in the diagnosis and management of neurological diseases. CSF is produced by plasma ultrafiltration and membrane secretion, is nearly free from cells, and protein levels are usually very low. However, CSF must be collected through a lumbar puncture, which is invasive and may be technically difficult in neonates. The quantity of fluid obtained may be limited and hemorrhage caused by the puncture may preclude interpretation. Thus, CSF metabolomics analyses have been performed in the experimental [21], but only rarely in the clinical setting of perinatal asphyxia [22].

3.2.4. Saliva

Although saliva can be collected non-invasively, the quantity may be limited. However, technological advances enable the analysis of small sample volumes (Table 2). Yen et al. [23] state that suctioning of the mouth typically yields 10–50 µL neonatal saliva, whereas sponges and wick applicators may yield slightly higher volumes, directly correlated to collection time. Oxidative stress markers glutathione [24], malondialdehyde [24–27] catalase, protein carbonyls, glutathione peroxidase, and 8-hydroxy-2′-deoxyguanosine [27], as well as isoprotanes, isofuranes, neuroprostanes, and neurofuranes [28] have all been measured in saliva from adults. However, although sporadically mentioned as an alternative [5,29], the evidence of metabolomics analysis of saliva in asphyxiated neonates is sparse, if not non-existing.

A selection of metabolites related to oxidative stress is presented in Table 2.

Table 2. Metabolites related to oxidative stress [29–32]. Reproduced with permission from Dr. Wishart, University of Alberta, Edmonton, Canada.

Metabolite	Description
Urea	The principal end product of protein catabolism.
Creatinine	A breakdown product of creatine phosphate in muscle.
Malonic acid	Malonic acid, also known as malonate or H2MALO, belongs to the class of organic compounds known as dicarboxylic acids and derivatives. In humans, malonic acid is involved in fatty acid biosynthesis
Methylguanidine	A guanidine compound deriving from protein catabolism. Synthesized from creatinine concomitant with the synthesis of hydrogen peroxide from endogenous substrates in peroxisomes. A nitric oxide synthase inhibitor.
L-Alanine	Alanine (Ala), also known as L-alanine, is an alpha-amino acid. Glutamate can transfer its amino group to pyruvate, a product of muscle glycolysis, through the action of alanine aminotransferase, forming alanine and alpha-ketoglutarate. Plasma alanine is often decreased when branched chain amino acids (BCAA) are deficient.
Succinate	Succinic acid (succinate) is a dicarboxylic acid.
Fumaric acid	Fumaric acid is a dicarboxylic acid.
Alpha-ketoglutarate	Oxoglutaric acid, also known as alpha-ketoglutarate, alpha-ketoglutaric acid, AKG, or 2-oxoglutaric acid. AKG is a nitrogen scavenger.
Hydroxycholesterol	27-Hydroxycholesterol (27-HC), also known as (25R)-cholest-5-ene-3β,26-diol or by its conventional name 26-hydroxycholesterol.
S-Adenosylmethionine	S-Adenosylmethionine, also known as SAM or acylcarnitine, belongs to the class of organic compounds known as 5′-deoxy-5′-thionucleosides. Possesses anti-inflammatory activity.
Glycine	An alpha-amino acid. Glycine is involved in the body's production of DNA, hemoglobin, and collagen, and in the release of energy.
Hypoxanthine	Hypoxanthine, also known as purine-6-ol or Hyp, belongs to the class of organic compounds known as purines. Under normal circumstances hypoxanthine is readily converted to uric acid.
Valine	Valine (Val) or L-valine is an alpha-amino acid. L-valine is a BCAA. The BCAAs consist of leucine, valine, and isoleucine (and occasionally threonine).
Choline	Important as a precursor of acetylcholine, as a methyl donor in various metabolic processes, and in lipid metabolism.
Glutathione	Like cysteine, glutathione contains the crucial thiol (-SH) group that makes it an effective antioxidant.
Ethylmalonate	Ethylmalonic acid, also known as alpha-carboxybutyric acid or ethylmalonate, is a branched fatty acid. Ethylmalonic acid can be synthesized from malonic acid.
3-Hydroxymethylglutaric acid	3-Hydroxymethylglutaric acid is an "off-product" intermediate in the leucine degradation process.
Glutaric acid	Is produced during the metabolism of some amino acids, including lysine and tryptophan.
Methylmalonic acid	Methylmalonic acid is a malonic acid derivative, which is a vital intermediate in the metabolism of fat and protein.

Table 2. *Cont.*

Metabolite	Description
Threonine	Threonine (Thr) or L-threonine is an alpha-amino acid. Threonine is sometimes considered a BCAA. Threonine is metabolized in at least two ways. In many animals it is converted to pyruvate via threonine dehydrogenase. An intermediate in this pathway can undergo thiolysis with CoA to produce acetyl-CoA and glycine.
3-Hydroxyisovaleric acid	A byproduct of the leucine degradation pathway.
Dimethylglycine	Dimethylglycine (DMG) is an amino acid derivative. The human body produces DMG when metabolizing choline into glycine. Homocysteine and betaine are converted to methionine and N,N-dimethylglycine by betaine-homocysteine methyltransferase.
Dimethylamine	An organic secondary amine.
Creatine	A naturally occurring non-protein compound classified as 'alpha amino acids and derivatives'. Its primary metabolic role is to combine with a phosphoryl group, via the enzyme creatine kinase, to generate phosphocreatine, which is used to regenerate ATP. It is naturally produced in the human body from the amino acids glycine and arginine, with an additional requirement for methionine to catalyze the transformation of guanidinoacetate to creatine.
Formic acid	The simplest carboxylic acid. Inhibition of cytochrome oxidase by formate may cause cell death by increased production of cytotoxic reactive oxygen species secondary to the blockade of the electron transport chain).

3.3. Animal Studies

In animal models of perinatal asphyxia, 100% oxygen was associated with metabolic markers of delayed cellular recovery and increased oxidative stress [30–32]:

In a piglet model of perinatal asphyxia, baseline urine metabolome, analyzed with ^{1}H NMR spectroscopy, distinguished between piglets that later became asystolic or not. The post-resuscitation-, seen in relation to the baseline metabolome, differentiated between piglets reoxygenated with different oxygen concentrations, primarily due to variations in metabolites with free radical scavenging properties: urea, creatinine, and malonate, as well as methylguanidine and hydroxyisobutyric acid [31]. A few years later, the same group investigated the same oxygen concentrations and confirmed distinct urine metabolomic patterns in piglets reoxygenated with 18, 21, 40, and 100% oxygen, respectively. Alanine and succinate were elevated, but glycine was unchanged in asphyxiated piglets resuscitated with 21% oxygen [32].

Moreover, in asphyxiated piglets and using flow injection analysis, MS/MS and LC-MS/MS, Solberg et al. [30] measured reduced blood succinate, fumarate, and alpha keto-glutarate indicating an earlier mitochondrial recovery when 21% oxygen was used for reoxygenation compared to 100% oxygen. Furthermore, oxysterols and acylcarnitine showed a differential response to 21% versus 100% oxygen reoxygenation. The ratios of alanine to branched chained amino acids, and of glycine to branched chained amino acids correlated with the duration of hypoxia.

Sachse et al. [11] reported a strong and consistent increase in alanine, succinate, hypoxanthine, and branched-chain amino acids in 125 asphyxiated piglets. There was a decrease of ^{1}H NMR signals associated with lipids. Baseline plasma hypoxanthine and lipoprotein concentrations were inversely correlated to the duration of hypoxia sustained before asystole occurred, but there was no evidence for a differential metabolic response to different resuscitation protocols including the use of supplementary oxygen, or in terms of survival [11]. In urine, branched-chain amino acids, especially valine, but also alanine and choline concentrations were higher after asphyxia and resuscitation compared to baseline.

3.4. Clinical Studies

Studies have comprehensively examined the metabolome in hypoxic-ischemic encephalopathy (HIE), but with a minor focus on oxygenation and oxidative stress [33,34]. A study of initial ventilation with 30% vs. 90% oxygen provided insights into oxidative stress markers in blood and urine of preterm infants $\leq$28 weeks of gestation, but not related to perinatal asphyxia [35].

3.4.1. Response to Hyperoxia

Vento et al. [36–39], showed that the blood reduced (GSH)-to-oxidized glutathione (GSSG) ratio was lower in asphyxiated infants that received 100% oxygen compared to 21% oxygen, lasting up to four weeks after birth [37]. Hundred percent oxygen ventilation was associated with increased activity of antioxidant enzymes including superoxide dismutase and glutathione redox cycle enzymes [38,39]. Urine N-acetyl-glucosaminidase correlated with GSSG and was significantly higher in infants that received 100% oxygen [36].

Dekker et al. [15] recently investigated initial ventilation with 30% versus 100% oxygen in preterm infants <30 weeks' gestation and found no difference in the lipid peroxidation product 8-iso-prostaglandin in umbilical cord blood, or infant blood at 1 and 24 h of age between 0% versus 100% oxygen.

3.4.2. Prognostic Utility

Negro et al. [40] measured blood advanced oxidation protein products (AOPP), non-protein bound iron (NPBI), and F2-isoprostanes (F2-IsoPs) in 84 infants with severe versus mild/moderate HIE at three different time points: P1 (4–6 h), P2 (24–72 h), and P3 (5 days). Mean (standard deviation) values of AOPP, NPBI, but not F2-IsoPs were significantly higher in infants with severe HIE with AOPP 34.1(39.2) vs. 15.7(15.5), $p = 0.033$ and NPBI 3.9 (4.4) versus 1.1(2.5), $p = 0.013$ at P1. However, there was no difference between groups at P2 and P3. A regression model showed that AOPP levels and male sex were both risk factors for higher brain damage scores; AOPP: OR 3.6, 95% confidence interval (1.1–12.2) and sex: OR 5.6, 95% confidence interval (1.2–25.7), respectively.

Vasiljevic et al. [22] performed routine lumbar puncture in 90 infants with HIE and reported a good relationship of glutathione peroxidase activity with the clinical stage of HIE ($p < 0.0001$) and gestational age ($p < 0.0001$). Glutathione peroxidase activity in CSF corresponded to later neurodevelopment outcome ($p < 0.001$) and showed a strong correlation with CSF levels of neuron-specific enolase ($p < 0.001$), a biomarker of the extent of brain injury.

Umbilical cord blood from term infants with confirmed HIE ($n = 31$) was compared to asphyxiated infants without encephalopathy ($n = 40$), and matched controls ($n = 71$) [41]. Targeted metabolomics revealed a significant increase in 29 metabolites from 3 distinct classes (amino acids, acylcarnitines, and glycerophospholipids) in infants with HIE or asphyxia compared to controls. Moreover, eight amino acids significantly increased in infants with HIE, but not asphyxia or matched controls. Thirteen acylcarnitines were significantly increased in infants with HIE and asphyxia without HIE, but the changes were more pronounced in the HIE population. A logistic regression model using five metabolites clearly delineated the severity of asphyxia and classified HIE infants with AUC = 0.92. These data suggest significant disruption to the energy pathways as well as nitrogen and lipid metabolism in both asphyxia and HIE.

Chu et al. [42] analyzed urine metabolite profiles from 256 asphyxiated infants and reported a positive relationship between suppressed biochemical networks involved in the macromolecular synthesis and perinatal asphyxia associated with significant oxidative stress and morbidity. In particular, elevated organic acids related to oxidative stress were significantly associated with neurodevelopmental outcomes: Ethylmalonate, 3-hydroxy-3-methylglutarate, 2-hydroxy-glutarate and 2-oxo-glutarate were associated with a good outcome; whereas glutarate, methylmalonate, 3-hydroxy-butyrate, and orotate were as-sociated with a poor outcome. In preterm asphyxiated infants, urine threonine and 3-hydroxyisovalerate were increased; and dimethylglycine, dimethylamine, creatine, suc-cinate, formate, urea, and aconitate were decreased [43]. These data demonstrated the potential application of bioinformatics methods in this metabolomics study and its potential clinical relevance.

Reinke et al. [44] proposed a metabolomic index ((succinate x glycerol) / (β-hydroxy-butyrate × O-phosphocholine)) to identify asphyxiated infants at risk of developing HIE, as elevated β-hydroxybutyrate, glycerol, O-phosphocholine, and succinate were most strongly associated with HIE severity. Ahearne et al. [45] later showed that if this index was <0.13, the outcome was likely to be normal (sensitivity 65% and specificity of 91%), whereas an index >2.4 was associated with a "severe outcome" (sensitivity of 80% and specificity of 100%). The authors concluded that this metabolomic index might be used in identifying neonates at risk of developing severe HIE.

4. Discussion

With a focus on oxidative stress, human and animal studies indicate that Krebs cycle intermediates including succinate, and hypoxanthine, in addition to acylcarnitines, play a role in perinatal asphyxia and may have prognostic and discriminatory properties.

The neonatal brain is highly oxygen dependent with low antioxidant capacity. This, combined with a high content of unsaturated fatty acids, makes the neonatal brain particularly vulnerable to oxidative stress [46]. Hypoxanthine is a purine metabolite, and levels are elevated during hypoxia. Hypoxanthine is oxidized in the presence of xanthine oxidase to uric acid during reoxygenation [47]. This generates an oxygen-free radical burst that, although uric acid has antioxidant properties, overwhelms endogenous antioxidant defense systems. Excessive production of reactive oxygen species [48] including hydrogen peroxide, hydroxyl free radical, superoxide anion radical, and reactive nitrogen species [49], proteases and caspases [50] in turn cause mitochondrial dysfunction and cell death. Free radical-induced lipid (arachidonic acid) peroxidation results in the generation of prostaglandin-like (prostanoids) isoprostanes [51,52]. Isoprostanes are often considered the gold standard for in vivo measurement of lipid peroxidation [53–55]. However, biofluid isoprostanes have demonstrated limited discriminatory properties in asphyxiated infants.

In asphyxia, incomplete fatty acid oxidation may result in an increase of fatty acid coenzyme A esters which bind to carnitine, resulting in the production of acylcarnitines [56]. High carnitine suppresses lipid peroxidation and the production of hydroxyl free radicals [57]. Carnitine thus has potent antioxidant activity [58], whereas acylcarnitines in perinatal asphyxia may indirectly reflect oxidative stress. Superoxide dismutase, glutathione, catalase, and glutathione peroxidase are cellular antioxidants [49]. Debuf et al. [29] performed a review of biomarkers of perinatal asphyxia and highlighted acylcarnitines as the most promising.

Several reviews, including one very recent [29], have addressed metabolomics biomarker analysis in perinatal asphyxia and HIE. However, they did not focus on oxidative stress the way our present review does. Perinatal asphyxia is associated with increased metabolites derived from lipid peroxidation. However, lipid peroxidation products seem to have limited utility in identifying infants at risk of developing HIE. Therefore, it has been proposed that the most reliable way to use metabolomics in perinatal asphyxia risk stratification and prognostication, is the establishment of a metabolic index composed of multiple metabolites that together have better prognostic value.

Biomarkers used for screening should be easy to collect, have a fast laboratory turnaround time, be reliable, and relatively inexpensive. Modern methods, e.g., LC-MS/MS have a turnaround time ranging from minutes to hours, opening for the possibility of integration into care. As stated by Debuf et al. [29], despite a current lack of point-of-care methods to measure these metabolites, considering the increasing advances in metabolomics, a "bench to the bedside" approach may be realistic within a reasonable time frame. Mussap et al. [5] suggested the translation of research results into the development of a low-cost device, e.g., a dipstick.

UPLC-MS/MS has been used to measure oxidative stress markers in saliva from adult patients and, although not supported by the available literature, saliva might be a suitable biofluid for point-of-care diagnostics. Saliva is, in theory, easy to collect and collection is non-invasive. However, saliva collection in newborns could be challenging, partially due to the limited amounts acquired. Standardized approaches to saliva collection, processing, and analyses are required for proper interpretation.

Strengths of this review include the innovative way of using existing knowledge to provide directions for research to develop non-invasive rapid tests of oxidative stress in perinatal asphyxia. Limitations include that we identified and report markers of oxidative stress in a wider sense and that we use indirect evidence, i.e., animal and adult data, to support our conclusions. However, we believe that the transparency of the reporting of results also leaves room for the readers to conclude from the limited knowledge base themselves.

In conclusion, in perinatal asphyxia, evidence about discriminatory metabolites related to oxidative stress, single or in combination, could be used to develop methods for rapid diagnostic and risk assessment purposes. Saliva might be a candidate biofluid for such point-of-care methods, however, more research is needed.

Author Contributions: A.L.S., S.N.Z., P.M.T. and G.M.S. performed data analysis and interpretation and drafted the initial version of the manuscript. All authors participated in critical revision of the manuscript for important intellectual content. All authors have read and agreed to the published version of the manuscript.

Funding: This research received no external funding.

Conflicts of Interest: The authors declare that the research was conducted in the absence of any commercial or financial relationships that could be construed as a potential conflict of interest.

References

1. DuPont, T.L.; Chalak, L.F.; Morriss, M.C.; Burchfield, P.J.; Christie, L.; Sanchez, P.J. Short-term outcomes of newborns with perinatal acidemia who are not eligible for systemic hypothermia therapy. *J. Pediatrics* **2013**, *162*, 35–41. [CrossRef] [PubMed]
2. Odd, D.E.; Lewis, G.; Whitelaw, A.; Gunnell, D. Resuscitation at birth and cognition at 8 years of age: A cohort study. *Lancet* **2009**, *373*, 1615–1622. [CrossRef]
3. Saugstad, O.D.; Oei, J.L.; Lakshminrusimha, S.; Vento, M. Oxygen therapy of the newborn from molecular understanding to clinical practice. *Pediatric Res.* **2019**, *85*, 20–29. [CrossRef]
4. Chalkias, A.; Fanos, V.; Noto, A.; Castren, M.; Gulati, A.; Svavarsdottir, H.; Iacovidou, N.; Xanthos, T. 1H NMR-metabolomics: Can they be a useful tool in our understanding of cardiac arrest? *Resuscitation* **2014**, *85*, 595–601. [CrossRef]
5. Mussap, M.; Antonucci, R.; Noto, A.; Fanos, V. The role of metabolomics in neonatal and pediatric laboratory medicine. *Clin. Chim. Acta* **2013**, *426*, 127–138. [CrossRef] [PubMed]
6. Sanchez-Illana, A.; Thayyil, S.; Montaldo, P.; Jenkins, D.; Quintas, G.; Oger, C.; Galano, J.M.; Vigor, C.; Durand, T.; Vento, M.; et al. Novel free-radical mediated lipid peroxidation biomarkers in newborn plasma. *Anal. Chim. Acta* **2017**, *996*, 88–97. [CrossRef] [PubMed]
7. Chattopadhyay, I.; Panda, M. Recent trends of saliva omics biomarkers for the diagnosis and treatment of oral cancer. *J. Oral Biosci.* **2019**, *61*, 84–94. [CrossRef]
8. Laputkova, G.; Schwartzova, V.; Banovcin, J.; Alexovic, M.; Sabo, J. Salivary Protein Roles in Oral Health and as Predictors of Caries Risk. *Open Life Sci.* **2018**, *13*, 174–200. [CrossRef] [PubMed]
9. Patil, D.J.; More, C.B. Salivary metabolomics—A diagnostic and biologic signature for oral cancer. *J. Oral Maxillofac. Surg. Med. Pathol.* **2021**, *33*, 546–554. [CrossRef]
10. Oldiges, M.; Lutz, S.; Pflug, S.; Schroer, K.; Stein, N.; Wiendahl, C. Metabolomics: Current state and evolving methodologies and tools. *Appl. Microbiol. Biotechnol.* **2007**, *76*, 495–511. [CrossRef]
11. Sachse, D.; Solevag, A.L.; Berg, J.P.; Nakstad, B. The Role of Plasma and Urine Metabolomics in Identifying New Biomarkers in Severe Newborn Asphyxia: A Study of Asphyxiated Newborn Pigs following Cardiopulmonary Resuscitation. *PLoS ONE* **2016**, *11*, e0161123. [CrossRef]
12. Ellis, D.I.; Goodacre, R. Metabolic fingerprinting in disease diagnosis: Biomedical applications of infrared and Raman spectroscopy. *Analyst* **2006**, *131*, 875–885. [CrossRef] [PubMed]
13. Krafft, C.; Steiner, G.; Beleites, C.; Salzer, R. Disease recognition by infrared and Raman spectroscopy. *J. Biophotonics* **2009**, *2*, 13–28. [CrossRef]
14. Cross, T.G.; Hornshaw, M.P. Can LC and LC-MS ever replace immunoassays? *J. Appl. Bioanal.* **2016**, *2*, 108–116. [CrossRef]
15. Dekker, J.; Martherus, T.; Lopriore, E.; Giera, M.; McGillick, E.V.; Hutten, J.; van Leuteren, R.W.; van Kaam, A.H.; Hooper, S.B.; Te Pas, A.B. The Effect of Initial High vs. Low FiO2 on Breathing Effort in Preterm Infants at Birth: A Randomized Controlled Trial. *Front. Pediatr.* **2019**, *7*, 504. [CrossRef] [PubMed]
16. Salvagno, G.L.; Danese, E.; Lippi, G. Preanalytical variables for liquid chromatography-mass spectrometry (LC-MS) analysis of human blood specimens. *Clin. Biochem.* **2017**, *50*, 582–586. [CrossRef]

17. Kong, S.T.; Lin, H.S.; Ching, J.; Ho, P.C. Evaluation of dried blood spots as sample matrix for gas chromatography/mass spectrometry based metabolomic profiling. *Anal. Chem.* **2011**, *83*, 4314–4318. [CrossRef] [PubMed]
18. Wilson, I. Global metabolic profiling (metabonomics/metabolomics) using dried blood spots: Advantages and pitfalls. *Bioanalysis* **2011**, *3*, 2255–2257. [CrossRef]
19. Michopoulos, F.; Theodoridis, G.; Smith, C.J.; Wilson, I.D. Metabolite profiles from dried blood spots for metabonomic studies using UPLC combined with orthogonal acceleration ToF-MS: Effects of different papers and sample storage stability. *Bioanalysis* **2011**, *3*, 2757–2767. [CrossRef] [PubMed]
20. Gika, H.G.; Theodoridis, G.A.; Wilson, I.D. Liquid chromatography and ultra-performance liquid chromatography-mass spectrometry fingerprinting of human urine: Sample stability under different handling and storage conditions for metabonomics studies. *J. Chromatogr. A* **2008**, *1189*, 314–322. [CrossRef]
21. Ventrella, D.; Laghi, L.; Barone, F.; Elmi, A.; Romagnoli, N.; Bacci, M.L. Age-Related 1H NMR Characterization of Cerebrospinal Fluid in Newborn and Young Healthy Piglets. *PLoS ONE* **2016**, *11*, e0157623. [CrossRef]
22. Vasiljevic, B.; Maglajlic-Djukic, S.; Gojnic, M.; Stankovic, S. The role of oxidative stress in perinatal hypoxic-ischemic brain injury. *Srp. Arh. za Celok. Lek.* **2012**, *140*, 35–41. [CrossRef]
23. Yen, E.; Kaneko-Tarui, T.; Maron, J.L. Technical Considerations and Protocol Optimization for Neonatal Salivary Biomarker Discovery and Analysis. *Front. Pediatr.* **2020**, *8*, 618553. [CrossRef] [PubMed]
24. Metgud, R.; Bajaj, S. Evaluation of salivary and serum lipid peroxidation, and glutathione in oral leukoplakia and oral squamous cell carcinoma. *J. Oral Sci.* **2014**, *56*, 135–142. [CrossRef] [PubMed]
25. Baltacioglu, E.; Yuva, P.; Aydin, G.; Alver, A.; Kahraman, C.; Karabulut, E.; Akalin, F.A. Lipid peroxidation levels and total oxidant/antioxidant status in serum and saliva from patients with chronic and aggressive periodontitis. Oxidative stress index: A new biomarker for periodontal disease? *J. Periodontol.* **2014**, *85*, 1432–1441. [CrossRef] [PubMed]
26. Nguyen, T.T.; Ngo, L.Q.; Promsudthi, A.; Surarit, R. Salivary Lipid Peroxidation in Patients With Generalized Chronic Periodontitis and Acute Coronary Syndrome. *J. Periodontol.* **2016**, *87*, 134–141. [CrossRef]
27. Tarboush, N.A.; Al Masoodi, O.; Al Bdour, S.; Sawair, F.; Hassona, Y. Antioxidant capacity and biomarkers of oxidative stress in saliva of khat-chewing patients: A case-control study. *Oral Surg. Oral Med. Oral Pathol. Oral Radiol.* **2019**, *127*, 49–54. [CrossRef] [PubMed]
28. Pena-Bautista, C.; Carrascosa-Marco, P.; Oger, C.; Vigor, C.; Galano, J.M.; Durand, T.; Baquero, M.; Lopez-Nogueroles, M.; Vento, M.; Garcia-Blanco, A.; et al. Validated analytical method to determine new salivary lipid peroxidation compounds as potential neurodegenerative biomarkers. *J. Pharm. Biomed. Anal.* **2019**, *164*, 742–749. [CrossRef]
29. Debuf, M.J.; Carkeek, K.; Piersigilli, F. A Metabolomic Approach in Search of Neurobiomarkers of Perinatal Asphyxia: A Review of the Current Literature. *Front. Pediatr.* **2021**, *9*, 674585. [CrossRef]
30. Solberg, R.; Enot, D.; Deigner, H.P.; Koal, T.; Scholl-Burgi, S.; Saugstad, O.D.; Keller, M. Metabolomic analyses of plasma reveals new insights into asphyxia and resuscitation in pigs. *PLoS ONE* **2010**, *5*, e9606. [CrossRef]
31. Atzori, L.; Xanthos, T.; Barberini, L.; Antonucci, R.; Murgia, F.; Lussu, M.; Aroni, F.; Varsami, M.; Papalois, A.; Lai, A.; et al. A metabolomic approach in an experimental model of hypoxia-reoxygenation in newborn piglets: Urine predicts outcome. *J. Matern. -Fetal Neonatal Med.* **2010**, *23* (Suppl. 3), 134–137. [CrossRef]
32. Fanos, V.; Noto, A.; Xanthos, T.; Lussu, M.; Murgia, F.; Barberini, L.; Finco, G.; d'Aloja, E.; Papalois, A.; Iacovidou, N.; et al. Metabolomics network characterization of resuscitation after normocapnic hypoxia in a newborn piglet model supports the hypothesis that room air is better. *BioMed Res. Int.* **2014**, *2014*, 731620. [CrossRef]
33. Pineiro-Ramos, J.D.; Cascant, M.M.; Nunez-Ramiro, A.; Lopez-Gonzalvez, A.; Solaz-Garcia, A.; Albiach-Delgado, A.; Martinez-Rodilla, J.; Llorens-Salvador, R.; Sanjuan-Herraez, D.; Quintas, G.; et al. Noninvasive monitoring of evolving urinary metabolic patterns in neonatal encephalopathy. *Pediatric Res.* **2021**, *127*, 49–54. [CrossRef]
34. Pineiro-Ramos, J.D.; Nunez-Ramiro, A.; Llorens-Salvador, R.; Parra-Llorca, A.; Sanchez-Illana, A.; Quintas, G.; Boronat-Gonzalez, N.; Martinez-Rodilla, J.; Kuligowski, J.; Vento, M.; et al. Metabolic Phenotypes of Hypoxic-Ischemic Encephalopathy with Normal vs. Pathologic Magnetic Resonance Imaging Outcomes. *Metabolites* **2020**, *10*, 109. [CrossRef] [PubMed]
35. Vento, M.; Moro, M.; Escrig, R.; Arruza, L.; Villar, G.; Izquierdo, I.; Roberts, L.J.; Arduini, A.; Escobar, J.J.; Sastre, J.; et al. Preterm Resuscitation With Low Oxygen Causes Less Oxidative Stress, Inflammation, and Chronic Lung Disease. *Pediatrics* **2009**, *124*, e439–e449. [CrossRef]
36. Vento, M.; Sastre, J.; Asensi, M.A.; Vina, J. Room-air resuscitation causes less damage to heart and kidney than 100% oxygen. *Am. J. Respir. Crit. Care Med.* **2005**, *172*, 1393–1398. [CrossRef] [PubMed]
37. Vento, M.; Asensi, M.; Sastre, J.; Garcia-Sala, F.; Pallardo, F.V.; Vina, J. Resuscitation with Room Air Instead of 100% Oxygen Prevents Oxidative Stress in Moderately Asphyxiated Term Neonates. *Pediatrics* **2001**, *107*, 642–647. [CrossRef]
38. Vento, M.; Asensi, M.; Sastre, J.; Lloret, A.; Garcia-Sala, F.; Minana, J.B.; Vina, J. Hyperoxemia caused by resuscitation with pure oxygen may alter intracellular redox status by increasing oxidized glutathione in asphyxiated newly born infants. *Semin. Perinatol.* **2002**, *26*, 406–410. [CrossRef] [PubMed]
39. Vento, M.; Asensi, M.; Sastre, J.; Lloret, A.; Garcia-Sala, F.; Viña, J. Oxidative stress in asphyxiated term infants resuscitated with 100% oxygen. *J. Pediatrics* **2003**, *142*, 240–246. [CrossRef]

40. Negro, S.; Benders, M.; Tataranno, M.L.; Coviello, C.; de Vries, L.S.; van Bel, F.; Groenendaal, F.; Longini, M.; Proietti, F.; Belvisi, E.; et al. Early Prediction of Hypoxic-Ischemic Brain Injury by a New Panel of Biomarkers in a Population of Term Newborns. *Oxid. Med. Cell. Longev.* **2018**, *2018*, 7608108. [CrossRef]
41. Walsh, B.H.; Broadhurst, D.I.; Mandal, R.; Wishart, D.S.; Boylan, G.B.; Kenny, L.C.; Murray, D.M. The metabolomic profile of umbilical cord blood in neonatal hypoxic ischaemic encephalopathy. *PLoS ONE* **2012**, *7*, e50520. [CrossRef] [PubMed]
42. Chu, C.Y.; Xiao, X.; Zhou, X.G.; Lau, T.K.; Rogers, M.S.; Fok, T.F.; Law, L.K.; Pang, C.P.; Wang, C.C. Metabolomic and bioinformatic analyses in asphyxiated neonates. *Clin. Biochem.* **2006**, *39*, 203–209. [CrossRef]
43. Longini, M.; Giglio, S.; Perrone, S.; Vivi, A.; Tassini, M.; Fanos, V.; Sarafidis, K.; Buonocore, G. Proton nuclear magnetic resonance spectroscopy of urine samples in preterm asphyctic newborn: A metabolomic approach. *Clin. Chim. Acta* **2015**, *444*, 250–256. [CrossRef]
44. Reinke, S.N.; Walsh, B.H.; Boylan, G.B.; Sykes, B.D.; Kenny, L.C.; Murray, D.M.; Broadhurst, D.I. 1H NMR derived metabolomic profile of neonatal asphyxia in umbilical cord serum: Implications for hypoxic ischemic encephalopathy. *J. Proteome Res.* **2013**, *12*, 4230–4239. [CrossRef]
45. Ahearne, C.E.; Denihan, N.M.; Walsh, B.H.; Reinke, S.N.; Kenny, L.C.; Boylan, G.B.; Broadhurst, D.I.; Murray, D.M. Early Cord Metabolite Index and Outcome in Perinatal Asphyxia and Hypoxic-Ischaemic Encephalopathy. *Neonatology* **2016**, *110*, 296–302. [CrossRef]
46. Ferriero, D.M. Neonatal brain injury. *N. Engl. J. Med.* **2004**, *351*, 1985–1995. [CrossRef] [PubMed]
47. Saugstad, O.D. Hypoxanthine as an indicator of hypoxia: Its role in health and disease through free radical production. *Pediatric Res.* **1988**, *23*, 143–150. [CrossRef] [PubMed]
48. Li, J.; Gao, X.; Qian, M.; Eaton, J.W. Mitochondrial metabolism underlies hyperoxic cell damage. *Free. Radic. Biol. Med.* **2004**, *36*, 1460–1470. [CrossRef]
49. Liu, J.; Litt, L.; Segal, M.R.; Kelly, M.J.; Pelton, J.G.; Kim, M. Metabolomics of oxidative stress in recent studies of endogenous and exogenously administered intermediate metabolites. *Int. J. Mol. Sci.* **2011**, *12*, 6469–6501. [CrossRef]
50. Fatemi, A.; Wilson, M.A.; Johnston, M.V. Hypoxic-ischemic encephalopathy in the term infant. *Clin. Perinatol.* **2009**, *36*, 835–858. [CrossRef]
51. Morrow, J.D.; Hill, K.E.; Burk, R.F.; Nammour, T.M.; Badr, K.F.; Roberts, L.J., 2nd. A series of prostaglandin F2-like compounds are produced in vivo in humans by a non-cyclooxygenase, free radical-catalyzed mechanism. *Proc. Natl. Acad. Sci. USA* **1990**, *87*, 9383–9387. [CrossRef] [PubMed]
52. Milne, G.L.; Yin, H.; Morrow, J.D. Human biochemistry of the isoprostane pathway. *J. Biol. Chem.* **2008**, *283*, 15533–15537. [CrossRef] [PubMed]
53. Montine, K.S.; Quinn, J.F.; Zhang, J.; Fessel, J.P.; Roberts, L.J., 2nd; Morrow, J.D.; Montine, T.J. Isoprostanes and related products of lipid peroxidation in neurodegenerative diseases. *Chem. Phys. Lipids* **2004**, *128*, 117–124. [CrossRef]
54. Niki, E.; Yoshida, Y.; Saito, Y.; Noguchi, N. Lipid peroxidation: Mechanisms, inhibition, and biological effects. *Biochem. Biophys. Res. Commun.* **2005**, *338*, 668–676. [CrossRef] [PubMed]
55. Tsimikas, S. In vivo markers of oxidative stress and therapeutic interventions. *Am. J. Cardiol.* **2008**, *101*, 34D–42D. [CrossRef]
56. Rebouche, C.J. Kinetics, pharmacokinetics, and regulation of L-carnitine and acetyl-L-carnitine metabolism. *Ann. N. Y. Acad Sci.* **2004**, *1033*, 30–41. [CrossRef]
57. Oka, T.; Itoi, T.; Terada, N.; Nakanishi, H.; Taguchi, R.; Hamaoka, K. Change in the membranous lipid composition accelerates lipid peroxidation in young rat hearts subjected to 2 weeks of hypoxia followed by hyperoxia. *Circ. J.* **2008**, *72*, 1359–1366. [CrossRef]
58. Reznick, A.Z.; Kagan, V.E.; Ramsey, R.; Tsuchiya, M.; Khwaja, S.; Serbinova, E.A.; Packer, L. Antiradical effects in L-propionyl carnitine protection of the heart against ischemia-reperfusion injury: The possible role of iron chelation. *Arch. Biochem. Biophys.* **1992**, *296*, 394–401. [CrossRef]

 *antioxidants*

Article

A Reductive Metabolic Switch Protects Infants with Transposition of Great Arteries Undergoing Atrial Septostomy against Oxidative Stress

José David Piñeiro-Ramos [1], Otto Rahkonen [2], Virpi Korpioja [3], Guillermo Quintás [4,5], Jaana Pihkala [2], Olli Pitkänen-Argillander [2], Paula Rautiainen [6], Sture Andersson [7], Julia Kuligowski [1,*] and Máximo Vento [1,8,*]

1. Neonatal Research Unit, Health Research Institute Hospital La Fe, Avenida Fernando Abril Martorell 106, 46026 Valencia, Spain; josedavidpineiro@gmail.com
2. Department of Paediatric Cardiology, New Children's Hospital, University of Helsinki and Helsinki University Hospital, Box 347, Stenbäckinkatu 9, 00029 Helsinki, HUS, Finland; Otto.Rahkonen@hus.fi (O.R.); jaana.pihkala@hus.fi (J.P.); Olli.Pitkanen@hus.fi (O.P.-A.)
3. Department of Children and Adolescents, Oulu University Hospital, P.O. Box 23, FIN-90029 OYS, 90570 Oulu, Finland; korpiojavirpi@gmail.com
4. Health & Biomedicine Unit, Leitat Technological Center, Par Cientific Barcelona, 08028 Barcelona, Spain; gquintas@leitat.org
5. Analytical Unit, Health Research Institute La Fe, Avenida, Fernando Abril Martorell 106, 46026 Valencia, Spain
6. Department of Anaesthesia and Intensive Care, New Children's Hospital, Helsinki University Hospital and University of Helsinki, Stenbackinkatu 9, 00029 Helsinki, Finland; Paula.Rautiainen@hus.fi
7. Pediatric Research Center, New Children's Hospital, Helsinki University Hospital and University of Helsinki, Stenbackinkatu 9, 00029 Helsinki, Finland; sture.andersson@hus.fi
8. Division of Neonatology, University & Polytechnic Hospital La Fe, Avenida Fernando Abril Martorell 106, 46026 Valencia, Spain
* Correspondence: julia.kuligowski@uv.es (J.K.); maximo.vento@uv.es (M.V.); Tel.: +34-96-1246661 (J.K.); +34-96-1246603 (M.V.)

Citation: Piñeiro-Ramos, J.D.; Rahkonen, O.; Korpioja, V.; Quintás, G.; Pihkala, J.; Pitkänen-Argillander, O.; Rautiainen, P.; Andersson, S.; Kuligowski, J.; Vento, M. A Reductive Metabolic Switch Protects Infants with Transposition of Great Arteries Undergoing Atrial Septostomy against Oxidative Stress. *Antioxidants* **2021**, *10*, 1502. https://doi.org/10.3390/antiox10101502

Academic Editor: Jacob V. Aranda

Received: 22 July 2021
Accepted: 16 September 2021
Published: 22 September 2021

Publisher's Note: MDPI stays neutral with regard to jurisdictional claims in published maps and institutional affiliations.

Abstract: Transposition of the great arteries (TGA) is one of the most common cyanotic congenital heart diseases requiring neonatal surgical intervention. Parallel circulations that result in impaired cerebral oxygen delivery already in utero may lead to brain damage and long-term neurodevelopmental delay. Balloon atrial septostomy (BAS) is often employed to mix deoxygenated and oxygenated blood at the atrial level. However, BAS causes a sudden increase in arterial blood oxygenation and oxidative stress. We studied changes in oxygen saturation as well as metabolic profiles of plasma samples from nine newborn infants suffering from TGA before and until 48 h after undergoing BAS. The plasma metabolome clearly changed over time and alterations of four metabolic pathways, including the pentose phosphate pathway, were linked to changes in the cerebral tissue oxygen extraction. In contrast, no changes in levels of lipid peroxidation biomarkers over time were observed. These observations suggest that metabolic adaptations buffer the free radical burst triggered by re-oxygenation, thereby avoiding structural damage at the macromolecular level. This study enhances our understanding of the complex response of infants with TGA to changes in oxygenation induced by BAS.

Keywords: transposition of the great arteries; balloon atrial septostomy; hypoxemia; metabolomics; oxidative stress; newborn; liquid chromatography-mass spectrometry (LC-MS)

1. Introduction

Transposition of the great arteries (TGA) is one of the most common cyanotic congenital heart diseases with an incidence of 0.3 per 1000 live births that requires surgical intervention in the neonatal period [1]. In hearts with TGA, systemic and pulmonary circulations run in parallel rather than in serial. This results in significant hypoxemia clinically reflected as central cyanosis. Survival after birth is only possible if there is an adequate blood mixing between the two circulations. Most hypoxemic neonates with TGA benefit from early institution of prostaglandin E1 (PGE) for ductal patency. If hypoxemia persists despite prostaglandin E1 (PGE1) infusion, balloon atrial septostomy (BAS) is needed to increase systemic oxygenation by improving the mixing of deoxygenated and oxygenated blood at the atrial level. After stabilization, arterial switch operation (ASO) is performed in the neonatal period. Mortality of ASO is low in the current era, however, morbidity is high and neonates with TGA are at risk of impaired neurodevelopmental outcome. Thus, long-term follow-up demonstrates that 30–50% of school-aged children with TGA show some form of developmental delay [2].

The underlying mechanism of developmental delay is thought to be multifactorial and include prenatal and postnatal factors. Hence, fetal hypoxia due to decreased oxygen delivery has been implicated in the abnormal brain development seen in newborns with TGA [3]. Moreover, reduced fetal cerebral oxygen consumption in TGA neonates has been associated with smaller head circumference and brain volume than those of normal neonates [4]. In addition, postnatal factors such as postnatal chronic hypoxemia, open-heart surgery with deep hypothermic circulatory arrest, and balloon atrial septostomy (BAS) have also been considered responsible for brain injury [5]. BAS improves mixing of systemic and pulmonary circulation and leads to an immediate increase in arterial oxygen content. However, BAS does not allow for full normalization of systemic oxygenation preoperatively. Very little is known about the direct effect of BAS on the neonatal brain, on cerebral oxygenation and oxygen metabolism [6], and whether the rapid increase of oxygen delivery results in brain reperfusion injury in neonates with TGA [7,8].

In mammals, aerobic metabolism with the concourse of oxygen is the most efficient biological means to supply energy required to sustain life. Under anaerobic conditions, pyruvate is converted into L-lactate. Anaerobic metabolism is by far less energy efficient than aerobic metabolism. Hence, in the absence of oxygen, the energy consumed by neurons rapidly leads to an exhaustion of the ATP reserves [9]. Due to the high metabolic rate of the brain, survival is almost exclusively dependent on the energy generated by aerobic glycolysis. The lack of oxygen stores and the reduced glycolytic capacity compel brain tissue to rely entirely on a continuous supply of oxygen and glucose provided by cerebral perfusion. Under these circumstances, acutely or chronically reduced oxygen availability due to environmental or pathophysiological causes inevitably leads to alterations of the brain structure and function [10].

Incomplete reduction of oxygen leads to the formation of reactive oxygen species (ROS), some of which are free radicals (e.g., anion superoxide and hydroxyl radicals). These extremely short half-life metabolites are capable of damaging nearby cellular components such as proteins, lipids, carbohydrates or DNA [11]. Both, acute and chronic hypoxia, enhance the formation of ROS through mitochondrial uncoupling provoking oxidative stress (OS) [12]. In addition, during reoxygenation, the increased availability of oxygen causes the activation of oxidases such as nicotinamide adenine dinucleotide phosphate (NADPH) oxidase or xanthine oxidase, further increasing the formation of anion superoxide and nitric oxide [11]. Neurons are highly vulnerable to the deleterious effects of ROS generated during acute hypoxia and/or hyperoxia. ROS trigger specific pathways that lead to apoptosis, necrosis, and inflammation of vulnerable areas of the brain causing long term neurodevelopmental, motor, and cognitive impairment [10].

Blood lactate has been largely employed as a surrogate for tissue hypoxia and/or ischemia. However, exclusive monitoring of serum lactate has neither provided sufficient insight into the magnitude of brain hypoxia nor conferred reliable prognostic information regarding long-term neurodevelopmental impairment [13,14]. More recently, comprehensive metabolic fingerprinting characterized by the simultaneous measurement of hundreds of metabolites from biological matrices has been increasingly employed for identifying predictive biomarkers or patient stratification [15].

In the present study we focused on the metabolic switch in infants with TGA after BAS. We performed serial analysis of lipid peroxidation byproducts as well as the plasma metabolome before and after BAS. This allowed us to study the impact of the rapid change in arterial blood oxygen content switching from a chronic hypoxic environment to an almost normoxic one, thus giving an insight into the dynamic hypoxia-related changes on the phenotypic level.

2. Materials and Methods

2.1. Study Population

We performed a prospective single center study to evaluate changes in cerebral oxygenation and metabolism before and for a period of 96 h following BAS in neonates with TGA. All patients with TGA admitted to Children's Hospital, Helsinki University Hospital, between 1 January 2015, and 1 June 2017, were considered for inclusion in the present study. Inclusion criteria included term gestational age and simple TGA without any significant associated heart defects (i.e., patients with ventricular septal defect were excluded). Reasons for failure to enroll included unavailability of parents for the consent process or parental refusal.

The study protocol was approved by the Ethics Committee of Helsinki University Hospital. All procedures were performed in accordance with relevant guidelines and regulations and written permission by signing an informed consent form or phone permission in urgent cases was obtained from legal representatives.

Data collected included peripheral oxygen saturation (SpO_2), mixed venous saturation, regional cerebral tissue oxygen saturation ($rcSO_2$) measured by near infrared spectroscopy (NIRS), heart rate, blood pressure, blood lactate levels, pH, base excess, and hemoglobin prior to and following BAS. Blood samples for metabolic analysis were collected from arterial cannula 5 min prior to and 5 min, 6 h, 24 h, 48 h, 72 h, and 96 h following BAS. At each timepoint, 1 mL of blood was collected into lithium heparin tubes, centrifuged, aliquoted, and stored at $-70\ ^\circ$C. Differences during $rcSO_2$, preductal peripheral oxygen saturation, fractional tissue oxygen extraction (FTOE), and cerebral oxygen extraction (CEO_2) were analyzed prior to and following BAS. Cerebral oxygen extraction was estimated from the difference of SaO_2 and ScO_2 as ScO_2 is close to venous SO_2. FTOE was calculated as CEO_2/SaO_2. Information regarding medications used prior to and following BAS was collected from electronic patient records.

2.2. Analytical Procedures

2.2.1. Lipid Peroxidation Biomarkers

Biomarkers of lipid peroxidation were analyzed in 63 plasma samples following previously published procedures [16,17]. Deuterated internal standards (IS) ($PGF_{2\alpha}$-d4 and 15-F_{2t}-Isoprostane-d4) were purchased from Cayman Chemical Company (Ann Arbor, MI, USA). For sample processing, 100 µL of plasma were thawed on ice and 100 µL of KOH solution at 15% (*w/v*) were added. The mixture was incubated at 40 $^\circ$C for 30 min. A volume of 3 µL of aqueous IS solution (20 µM) was added to hydrolyzed samples and diluted to 900 µL with H_2O:MeOH (85:15, 2.8% *v/v* HCOOH) solution. Then, the samples were mixed for 30 s at maximum speed and centrifuged at 16,000$\times$ *g* and 4 $^\circ$C for 10 min. For clean-up and pre-concentration of the samples, an SPE procedure employing Discovery® DSC-18 SPE 96-well plates from Sigma Aldrich Química S.A (Madrid, Spain) was carried out. First, the stationary phase was equilibrated with 1 mL of MeOH and 1 mL of water. Then, the supernatant of the centrifuged and diluted sample was loaded followed by washing with 1 mL of H_2O (0.1% *v/v* HCOOH, pH 3) and 500 mL heptane. Finally, cartridges were dried with room air and the compounds of interest were eluted with 4 $\times$ 100 µL ethyl acetate. The eluate was evaporated using a miVac centrifugal vacuum concentrator (Genevac LTD, Ipswich, UK) and dissolved in 60 µL H_2O (0.1% *v/v* HCOOH, pH 3):CH_3OH (85:15 *v/v*).

An Acquity-Xevo TQS system from Waters (Milford, MA, USA) operating in negative electrospray ionization (ESI$^-$) mode was employed for UPLC-MS/MS analysis. A Waters BEH C$_{18}$ column (2.1 mm $\times$ 100 mm, 1.7 µm, Waters, Wexford, Ireland) was used. Flow rate, column temperature, and injection volume were set at 450 µL min^{-1}, 45 °C, and 9 µL, respectively. A binary mobile phase H$_2$O (0.1% *v/v* HCOOH):CH$_3$CN (0.1% *v/v* HCOOH) gradient with a total runtime of 7.0 min was run as follows: from 0.0 to 0.1 min 15% *v/v* CH$_3$CN (0.1% *v/v* HCOOH) (mobile phase channel B); from 0.1 to 5.0 min % B increased up to 40%; from 5.0 to 6.0 min % B increased up to 75%; between 6.0 and 6.15 conditions were held constant at 75% B followed by the return to initial conditions (i.e., 15% B) between 6.15 and 6.25 min; conditions were maintained for 0.75 min for system re-equilibration. ESI interface conditions were selected as follows: capillary voltage was set to 2.9 kW; source and desolvation temperatures were 150 °C and 395 °C, respectively; and nitrogen cone and desolvation gas flows were 150 and 800 L h^{-1}, respectively. Parameters selected for determination of lipid peroxidation biomarkers are shown in Table 1.

Table 1. Mass spectrometric parameters and chromatographic windows employed for the lipid peroxidation biomarkers.

Analyte [p.d.u.]	RT [min]	Parent Ion (*m/z*)	Daughter Ion (*m/z*)	CE (eV)	Cone Voltage (V)
IsoPs	4.3–6.6	353.20	115.00	30	35
Di-homo-IsoPs	5.0–6.8	381.00	143.00	20	20
Di-homo-IsoFs	3.5–6.5	397.00	155.00	24	35
NeuroFs	2.70–6.50	393.00	193.00	20	35
IsoFs	2.1–6.60	369.20	115.00	20	45
NeuroPs	2.30–6.50	377.00	101.00	20	35

Note: IsoPs = isoprostanes. IsoFs = isofurans. NeuroFs = neurofurans. NeuroPs = neuroprostanes. RT stands of Retention times. CE stands of collision energy.

2.2.2. Untargeted Ultra-Performance Liquid Chromatography Coupled to Time-of-Flight Mass Spectrometry (UPLC-TOFMS) Metabolomic Analysis

Plasma samples were thawed on ice and homogenized on a Vortex mixer. 75 µL of cold acetonitrile were added to 25 µL of plasma, homogenized and kept on ice during 15 min followed by centrifugation at 16,000$\times$ *g* during 15 min at 4 °C. 80 µL of supernatant were collected and transferred to a 96 well plate, evaporated to dryness on a miVac centrifugal vacuum concentrator (Genevac LTD, Ipswich, UK) at room temperature and dissolved in 60 µL of an internal standard (IS) solution containing betaine-D$_{11}$, methionine-D$_3$, hypoxanthine-D$_3$, cystine-D$_4$, tyrosine-D$_2$, prostaglandinF$_{2\alpha}$-D$_4$, uridine-C^{13}N^{15}, reserpine, phenylalanine-D$_5$, leucine enkephalin, caffeine-D$_9$, and tryptophane-D$_5$ with purities $\geq$ 99% at a concentration of 1.5 µM in H$_2$O:CH$_3$CN (0.1% HCOOH) (95:5 *v/v*).

A QC sample was prepared by mixing 5 µL of each plasma sample and a total of three aliquots were processed alongside with the plasma samples applying the same procedures. A blank extract was prepared by using a heparinized syringe and 0.5 mL of ultrapure H$_2$O and processed as described for plasma samples.

For chromatographic separations, an Agilent Technologies (Santa Clara, CA, USA) 1290 Infinity UPLC chromatograph equipped with a UPLC ACQUITY BEH C18 column (2.1 mm $\times$ 100 mm, 1.7 µm, Waters, Wexford, Ireland) was employed. Autosampler and column temperatures were set to 4 and 40 °C, respectively. A flow rate of 400 µL min^{-1} and an injection volume of 4 µL were used. Separations were carried out keeping 98% of mobile phase A (H$_2$O, 0.1% *v/v* HCOOH) for 0.5 min, followed by a linear gradient from 2 to 20% of mobile phase B (CH$_3$CN, 0.1% *v/v* HCOOH) in 3.5 min and from 20 to 95% B in 4 min. Conditions of 95% B were maintained for 1 min and a 0.25 min gradient was used to return to the initial conditions, which were held until reaching 8.5 min.

Full-scan MS data were acquired between 100 and 1700 *m/z* with a scan frequency of 6 Hz (1274 transients/spectrum) on an iFunnel quadrupole time-of-flight (QTOF) Agilent 6550 spectrometer operating in the TOF MS mode. The following electrospray ionization settings were used: gas T, 200 °C; drying gas, 14 L min^{-1}; nebulizer, 37 psig; sheath gas

T, 350 °C; sheath gas flow, 11 L min^{-1}. A mass reference standard used for automatic MS spectra re-calibration during analysis was introduced into the source via a reference sprayer valve using the 149.02332 (background contaminant), 121.050873 (purine), and 922.009798 (HP-0921) *m/z*. MassHunter workstation from Agilent was employed for data acquisition and manual integration of ISs.

Before launching the analytical sequence, system suitability was checked employing a standard mixture containing ISs. The analytical system was conditioned by eight repeated injections of the QC at the beginning of the batch. Data acquired during system conditioning were discarded from data analysis. A total of 63 plasma sample extracts were analyzed in randomized order in a single analytical batch using the positive electrospray (ESI$^+$) mode. QC samples were analyzed every 6th sample and at the beginning and end of the batch for assessment and correction of instrumental performance [18]. The blank extract was injected a total of two times (once during system conditioning and once at the end of the batch) and used for data clean-up with the aim of identifying signals from other than biological origin. Subsequently, sample analysis was carried out in ESI$^-$ mode repeating the same protocol described for the ESI$^+$ mode.

2.2.3. Data Processing and Statistical Analysis

ProteoWizard [19] (http://proteowizard.sourceforge.net (accessed on 20 September 2021)) software was used for conversion of raw UPLC-TOFMS data into centroid mzXML format. A peak table was extracted using XCMS (version 3.4.2) [20–22] (https://bioconductor.org/packages/release/bioc/html/xcms.html (accessed on 20 September 2021)) running in R (version 3.5). For peak detection, the centWave method was used as follows: ppm = 20, peakwidth = (4 and 25), snthresh = 10. For the resolution of overlapping peaks, a minimum *m/z* difference of 7.5 mDa was selected. For each extracted feature, the 'wMean' function was used for calculating intensity-weighted *m/z* values and peak limits used for integration were found through descent on the Mexican hat filtered data. For peak grouping the "nearest" method with mzVsRT = 1 and retention time (RT) and *m/z* tolerances of 6 s and 10 mDa, respectively, was used. Missing peak data was filled applying the fillPeaks method with the default parameters. A total of 18,582 and 13,479 features were initially detected after peak detection, integration, chromatographic deconvolution, and alignment in ESI$^+$ and ESI$^-$ modes, respectively. The CAMERA [23] package was used for identifying peak groups and annotation of isotopes and adducts using the following settings: sigma = 6, perfwhm = 0.5, ppm = 20.

Peak integration accuracy was checked by comparing the generated peak table with areas obtained from manual integration of ISs. Peak intensities of ISs and QC samples were used for assessing the instrumental response during data acquisition throughout the batch as described elsewhere [24,25]. The Quality Control-Support Vector Regression (QC-SVR) algorithm [26] and the LIBSVM library [27] were used for correcting intra-batch variation using an ε–range of 2.5 to 7.5 and a γ-range of 1 to 10^5. C was defined for each feature as the median value in QCs. Then, features detected in blanks (<5× signal of the blank) and those with an RSD% in QC samples ≥20% were excluded. The final peak tables contained 3886 and 5600 features for ESI$^+$ and ESI$^-$, respectively, and were searched for molecular ion peaks of drugs and known drug metabolites that have been administered to infants, their isotope as well as Na and K adducts (*m/z* tolerance: 10 mDa). Metabolic features that were identified as drugs or their metabolites, isotopes and adducts were excluded from further data analysis.

MATLAB 2019b inbuilt functions as well as in-house written scripts (available from the authors upon reasonable request) and the PLS Toolbox 8.0 from Eigenvector Research Inc. (Wenatchee, WA, USA) were used for Principal Component Analysis (PCA) and the computation of Pearson correlations. For PCA, data sets generated in ESI$^+$ and ESI$^-$ were concatenated. MetaboAnalyst (version 4.0) [28] was used for hierarchical clustering and the generation of heatmaps employing Euclidean distance and Ward's method (statistical analysis tool). Pathway analysis was carried out using MetaboAnalyst with the MS peaks to pathways tool (mass accuracy = 10 ppm, mummichog algorithm with top 10% peaks *p*-value cut-off) in the 4-column format (*m/z*, RT, ionization mode, *p*-value of Pearson correlation between metabolic features and FTOE) and the Kyoto Encyclopedia of Genes and Genomes (KEGG) pathway library (*Homo sapiens*). Metabolomics data are available on Zenodo (https://zenodo.org/record/4495124#.YBpwoC1DkWp (2 February 2021)).

The non-parametric Wilcoxon rank-sum test was used for assessing changes in levels of biomarkers of lipid peroxidation over time. *p*-values from Spearman correlations were used for clinical data, where appropriate.

3. Results

3.1. Clinical Results

A total of 12 newborn infants fulfilled the study requirements. Out of these, one patient died, and two patients did not require BAS. The remaining 9 patients who underwent BAS, completed all analysis. Demographics and perinatal characteristics are detailed in Table 2. Six patients (50%) had prenatal diagnosis and nine patients (75%) underwent BAS due to low preductal saturation at the age 4.6 ($\pm$ 2.7) hours. There was one early death prior to BAS. The patient was transferred from another central hospital with prostaglandin infusion but had extremely low preductal saturations (<30%), severe lactate acidosis, and was in pulseless electrical activity at the time of admission and care was withdrawn. Figure 1 describes evolving SpO_2, $rcSO_2$, FTOE, and CEO_2 before and after BAS. In the nine patients included in the study, the lowest preductal peripheral oxygen saturation (SpO_2) at admission had a median of 64.5% (range 39.0–92.0). Preductal SpO_2 increased from a median of 85.6% (range 62.0–90.6) before BAS to 89.1% (range 81.8–93.5) 6 h following BAS and 90.0% (range 85.2–93.6) 24 h following BAS. $rcSO_2$ at the same time points were 50.0% (range 35.0–70.0), 52.8% (range 36.4–72.5), 63.0% (range 48.2–74.1), and 69.2 (range 58.8–80.8), respectively. $rcSO_2$ correlated strongly with simultaneously measured SpO_2 (Spearman's r = 0.89, *p*-value < 0.001). CEO_2 increased after BAS (27.2–28.1) but both, CEO_2 and FTOE, decreased 24 h following BAS (see Figure 1). Complete recovery of cerebral oxygen saturation did not occur until 24 h after BAS.

Table 2. Patients' demographics and timing of postnatal clinical and analytical interventions.

Patients' Demographics	All Patients	BAS Patients
Patients recruited, N (%)	12 (100)	9 (100)
Gender, N (%)		
- Male	10 (83)	8 (89)
- Female	2 (17)	1 (11)
Balloon atrial septostomy, N (%)	9 (75)	9 (100)
Expired, N (%)	1 (8.3)	0 (0)
Prenatal diagnosis, N (%)	6 (50)	4 (44)
Gestational age, weeks (median, range)	39.5 (37.0–41.3)	39.9 (38.4–41.3)
Birth weight, kg (median, range)	3.4 (2.0–4.0)	3.5 (3.1–4.0)
Postnatal age, hours (median, range)		
- At the time of the first measured peripheral saturation	1.7 (0.5–13.2)	2.3 (0.5–6.7)
- At time of BAS		3.2 (2.0–9.1)
Saturation, % (median, range)		
- First peripheral saturation (SpO_2)	79 (51–91)	77 (56–88)
- Lowest peripheral saturation (SpO_2)	65 (39–86)	65 (39–86)
- First regional cerebral oxygen saturation ($rcSO_2$)	58 (37–79)	53 (37–79)
- 5-min post septostomy peripheral saturation (SpO_2)	86 (62–91)	86 (62–91)
Blood lactate, mmol/L (median, range)		
- Before BAS	3.2 (1.3–4.9)	3.6 (1.3–4.9)
- 5-min after BAS	2.7 (1.4–5.0)	3.3 (1.4–5.0)

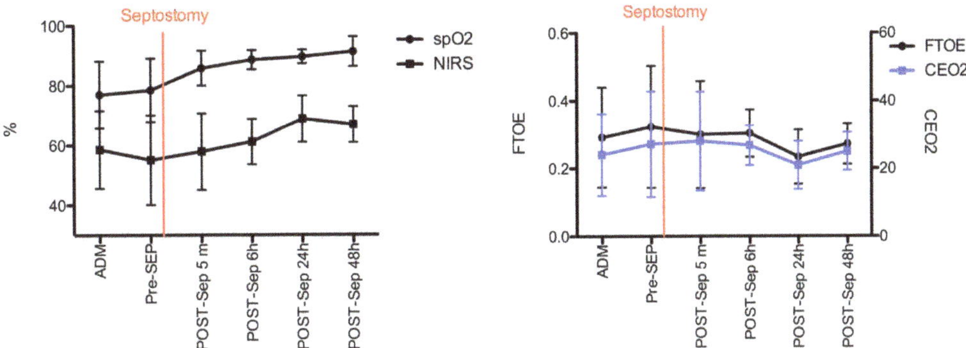

Figure 1. Evolution of SpO2, rcSO2 (**left**), FTOE, and CEO2 levels (**right**) before and after BAS. Black and blue lines and error bars are median and 25th and 75th percentile, respectively.

3.2. Lipid Peroxidation Biomarkers

Total di-homo-isoprostanes, di-homo-isofurans, and isoprostanes were excluded as these parameters were found <LOQ in all study samples. Figure 2 depicts relative responses of total neurofurans, isofurans, and neuroprostanes obtained over time. No significative changes over time were detectable (Wilcoxon rank sum test, p-values > 0.05). Furthermore, no strong (Pearson correlation coefficients > |0.5|) and significant correlations were found between isoprostanoid levels and FTOE.

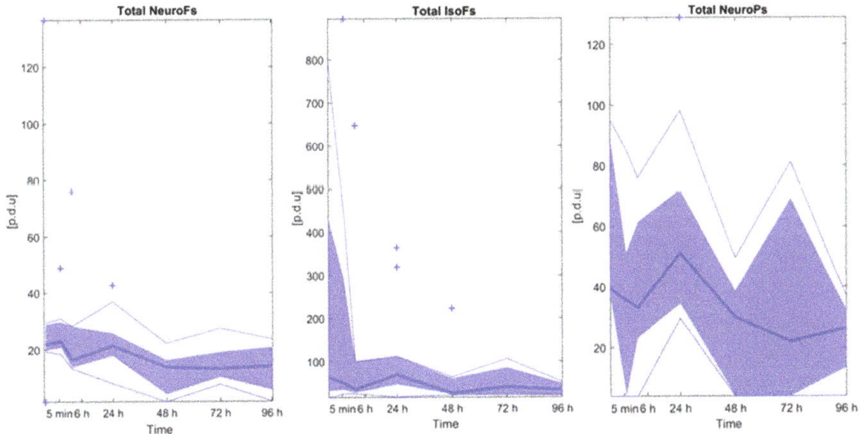

Figure 2. Isoprostanoid levels over time. Note: Bold blue lines correspond to median values, blue areas correspond to the interquartile range (1st and 3rd quartile), thin blue lines correspond to minimum and maximum values, and + correspond to outliers. Note: NeuroFs = neurofurans; IsoFs = isofurans; NeuroPs = neuroprostanes.

3.3. Effect of Time on the Plasma Metabolome

Figure 3 shows a PCA scores plot of the plasma metabolic fingerprint of infants before BAS and at different time points after BAS. The scores plot from PC1 vs. PC2 illustrates the impact of the sampling time point on the plasma metabolome. For most patients, a time-dependent shift towards lower scores on PC1 with increasing time after septostomy was observed. Even though the applied data analysis workflow included the removal of drug metabolites, this effect might, at least partially, be related to the employed medication and the procedure itself. Also, a high inter-individual variation can be noted.

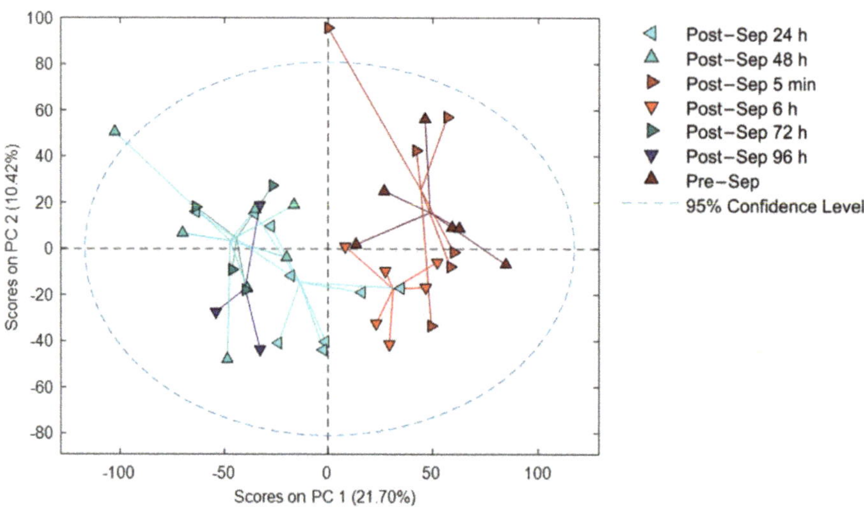

Figure 3. PCA results of samples collected before and at different time points after BAS. Note: Sep stands for septostomy.

3.4. Correlation of Metabolic Features with FTOE

FTOE reflects the balance between oxygen supply and consumption in tissue and can, therefore, be used as an indicator of inadequate tissue perfusion and oxygenation. We specifically focused on modelling the effects of metabolic changes in the plasma metabolome associated with FTOE. Pearson correlations for each metabolic feature with FTOE before and at different time points post-septostomy were calculated as shown in Figure 4. A panel of features with a significant association with FTOE (i.e., p-values from Pearson correlation < 0.05) was identified including both, positively (red dots, correlation coefficient > 0.5) and negatively (blue dots, correlation coefficient < -0.5) correlated features. Figure 5 shows a heatmap of the relative intensities of significantly correlated features. Two distinct clusters can be observed, with plasma fingerprints from samples collected before and 5 min as well as 6 h after BAS belonging to one cluster, and samples collected 24–96 h after BAS belonging to a second cluster. Most metabolic features in Figure 5 showed a decreasing trend in relative intensities when comparing cluster one to cluster two. Pathway analysis detected four significantly altered pathways associated with changing FTOE in infants with TGA undergoing BAS (see Table 3).

Table 3. Pathway alterations associated with changing FTOE in infants with TGA undergoing atrial septostomy.

Pathway Name	Compound Code (KEGG)	Pathway ID	# Hits	# Sig Hits	p-Value
Pentose phosphate pathway	C01801; C00672; C00121; C00121; C00257; C00257; C00258	map00030	7	7	0.00002
Pentose and glucuronate interconversions	C01068; C00181; C00259; C00310; C00312; C00379; C00532; C00379; C00532; C00379; C00379; C00532; C00257; C00257; C00029; C00052; C00191; C00618; C02266	map00040	11	7	0.003
Ascorbate and aldarate metabolism	C00137; C00029; C02670; C00191; C00800	map00053	8	5	0.013
Inositol phosphate metabolism	C00137; C00222; C00191	map00562	4	3	0.03

Note: Mummichog input: 10 ppm; p-value cut-off: 10%; KEGG database. #: stands for number (Number of hits, number of significant hits).

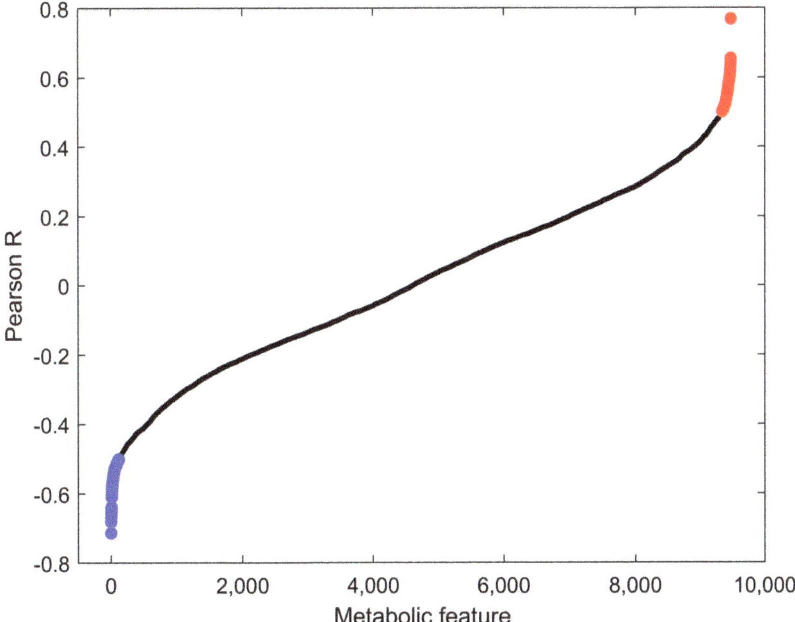

Figure 4. Pearson correlation coefficients between metabolic features and FTOE. Blue: features with *p*-values < 0.05 and correlation coefficients < −0.5; red: features with *p*-values < 0.05 and correlation coefficients > 0.5; black: all remaining features.

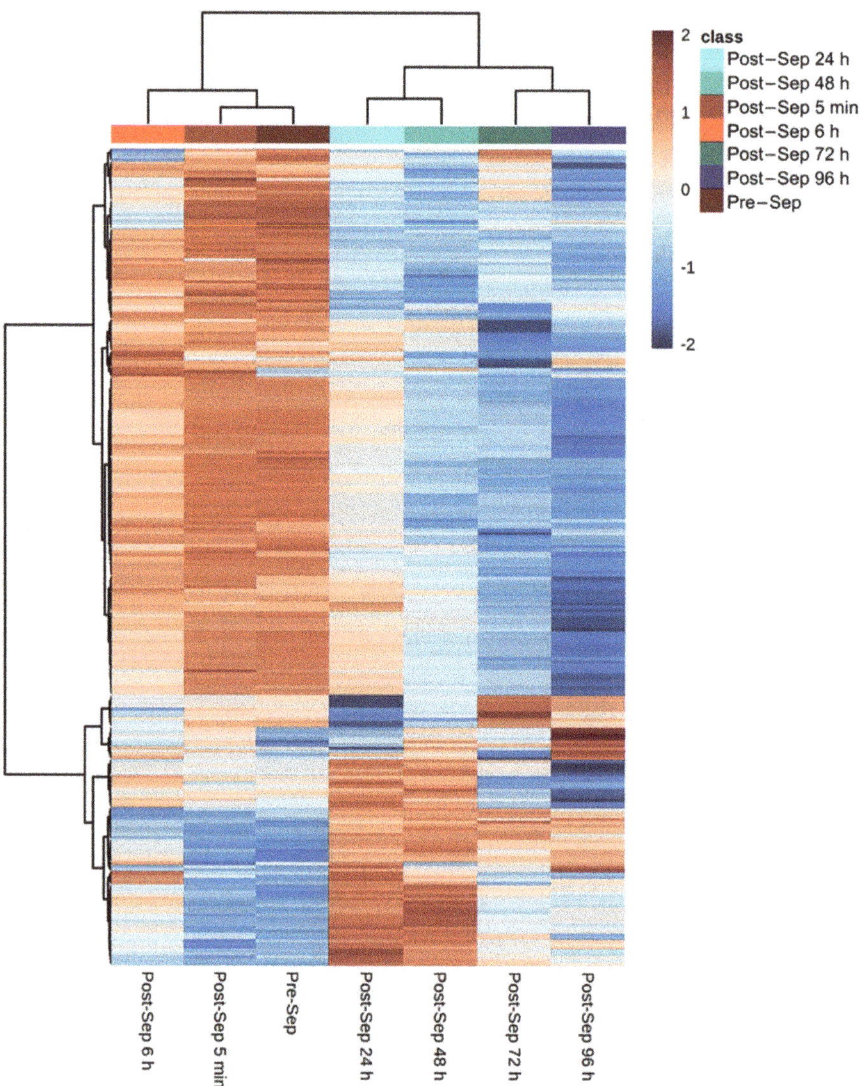

Figure 5. Relative intensities of metabolic features that correlate (i.e., *p*-value < 0.05 and |correlation coefficients| > 0.5) with FTOE. Note: color represents autoscaled relative intensities; blue—metabolite levels lower than average, and red—metabolite levels higher than average. *p*-values from Pearson correlation were computed and significantly correlating metabolites are shown; heatmap calculated using Euclidean distance and Ward algorithm.

4. Discussion

TGA is a severe congenital cardiac malformation that causes hypoxemia during fetal life and in the newborn period [1]. TGA has deleterious consequences on growth and development due to a deficient tissue oxygenation that especially affects the Central Nervous System [2,4]. We report the first metabolomic study involving neonates with TGA who underwent BAS. We performed a study on oxidative stress biomarkers as well as a comprehensive qualitative characterization of plasma metabolites before and after BAS.

BAS in TGA patients caused a rapid switch from hypoxia to normoxia. Preductal peripheral pulse oximetry saturation and rcSCO$_2$ increased rapidly after BAS while simultaneously FTOE decreased. As a result, brain oxygenation substantially improved (See Figure 1). Thereafter, changes in oxygenation plateaued. Complete recovery of cerebral oxygen saturation occurred only 24 h after BAS. A gradual change was observed in the metabolome, accordingly (See Figure 3). The provision of energy to satisfy metabolic demands of the brain is exclusively dependent on aerobic metabolism and oxygen deprivation caused by hypoxia and/or ischemia for just a few minutes may cause severe brain damage. Mitochondrial ROS production is regulated through tissue succinate levels and the activity of oxidases (NADPH oxidase, xanthine oxidase) [29]. During hypoxia and ulterior reoxygenation both, succinate levels and oxidase activation, generate a burst of ROS that directly damage tissue structure and function [29]. Moreover, the pro-oxidant imbalance provokes the activation of the caspase pathway and of transcription factor NFkB. Subsequently, programmed cell death and inflammation are triggered for hours, days, or even weeks. Consequently, there is an amplification of the initial area of brain injury that contributes to aggravate long-term neurological prognosis [30–32].

Isoprostanes and isofurans, but especially neuroprostanes and neurofurans are highly sensitive to oxidative stress related brain damage [33]. However, we did not observe changes in those compounds after BAS (See Figure 2). The attenuation of the expected pro-oxidant status after BAS suggests a pattern of metabolomic changes with a reducing profile. In this regard, post-BAS untargeted metabolomics evidenced a significant enhancement in the activity of the pentose phosphate pathway (PPP). The central role of the PPP has attracted more attention in recent years. Emerging evidence suggests that the PPP is tightly and meticulously controlled in cells and that its abnormal regulation leads to uncontrolled redox homeostasis [34]. The PPP has shown great versatility for *de novo* nucleotide biosynthesis via ribose-5-phosphate and adopts a simultaneous organization with glycolysis to produce NADPH. Nucleotide biosynthesis possibly participates in DNA damage repair. Our pathway analysis data revealed changes in the relative concentrations of 2-deoxy-D-ribose 1-phosphate (C00672), a precursor of ribose-5-phospate, and substrate of phosphoribosyl pyrophosphate (PRPP). This substrate is an essential compound of purine, pyridine, and histidine synthesis. NAPDH is a cofactor of glutathione reductase, an essential enzyme in the glutathione redox cycle that contributes to the reconversion of GSSG into GSH thus contributing to the normalization of the GSH/GSSG pair and the removal of ROS [35,36]. In this context, the pathway analysis data suggest an alteration of gluconic acid (C00257). The phosphorylation of this compound generates 6-phosphogluconate, an essential substrate in the oxidative branch of the PPP. Furthermore, alterations of other compounds such as ribitol-5-phosphate (C01068) and 2-deoxy-D-ribose (C00672) contribute as substrates for key PPP compounds. Indeed, the activity of the PPP is rapidly re-routed when cells are exposed to an oxidative burst. This response is exquisitely adjusted by cooperating of metabolic and gene regulatory mechanisms. Metabolic changes imply the inactivation of glycolytic enzymes which occurs immediately after the oxidant aggression thus blocking glycolysis [37]. Thereafter, the transcriptional response takes over and maintains higher PPP activity through up-regulation of enzymes and post-translational modifications including those which increase the activity of G6PDH [38]. Furthermore, to counteract mitochondrial ROS production under normal metabolic circumstances, but also during hypoxia or hypoxia-reoxygenation, steady NADPH production becomes essential as it represents the main electron donor for the generation of GSH that will provide electrons for the reduction of detrimental peroxides by glutathione peroxidase [34].

In addition, our results reveal the alteration of the pentose and glucuronate interconversion pathway, another key pathway in the homeostasis of metabolic pathways. The glucuronate pathway is an alternative pathway for the oxidative degradation of glucose without the production of ATP. Substrate compounds such as xylose (C00181), arabinose (C00259), arabitol (C00532), xylulose (C00312), UDP-glucose (C00029), glucuronic acid (C00191), dehydrogulonate (C00618), xylonolactone (C02266), and ribitol-5-phosphate

(C01068) were altered according to the reported pathway analysis data. Interestingly, in humans the synthesis of ascorbic acid is not feasible, therefore a substantial proportion of uridine diphosphate glucoronate (UDP-glucoronate) is converted into xylulose-5-phosphate which is further metabolized through the PPP to fuel NADPH production and promote the preservation of a reduced environment [39].

Finally, pathway analysis also showed an alteration of ascorbate and aldarate and inositol phosphate metabolism. The ability of ascorbate to donate electrons enables it to act as a free radical scavenger and to reduce higher oxidation states of iron to Fe^{2+}. Ascorbic acid is an important antioxidant in plasma, where it consumes oxygen free radicals. Erythrocytes have a high capacity to regenerate ascorbate from its two electron-oxidized form, dehydroascorbic acid. Intracellular dehydroascorbic acid is rapidly reduced to ascorbate by GSH in a direct chemical reaction, or indirectly with the concurrent action of glutaredoxin and thioredoxin reductases. Intracellular ascorbate can spare, and possibly recycle, alpha distocopherol in the erythrocyte membrane. In turn, alpha tocopherol protects the cell membrane from lipid peroxidation. The ability of erythrocytes to recycle ascorbate, coupled with the ability of ascorbate to protect alpha tocopherol in the cell membrane and in lipoproteins, provides a potentially important mechanism for preventing lipid peroxidative damage secondary to hypoxia or hypoxia-reoxygenation events [40].

We acknowledge limitations of our study. First, the number of subjects included is limited and some of the blood samples during BAS were not collected. We would like to stress the stringent inclusion criteria applied during patient recruitment and the low incidence of the condition. The present data provides evidence to justify large multi-center efforts for validating the current findings. Finally, we lack a control group of healthy infants for obvious ethical reasons.

5. Conclusions

In summary, this is a comprehensive metabolomic assessment of neonates with TGA. The results obtained suggest differences in oxygen supply and consumption in cerebral tissue during hypoxia and near-normoxia. The number of patients is limited, but the combined assessment of lipid peroxidation biomarkers and untargeted metabolomic screening of a cohort of infants with TGA undergoing BAS provides insightful information to understand the physiopathology of this complex disease. The metabolic switch after BAS causes oxidative stress. However, oxidative stress may at least be partially neutralized by the induction of different metabolic pathways but especially the PPP that supplies with reductive electrons. From a clinical point of view, although supplemental arterial oxygenation has limited effects on oxygenation in parallel circulation, our results suggest potential benefits of avoiding hyperoxia in patients undergoing BAS to prevent from attenuating the antioxidant effect inherent to the metabolic switch after septostomy.

Author Contributions: Conceptualization, O.R., M.V., and J.K.; Methodology, G.Q., J.K., O.R., and S.A.; Software, G.Q.; Validation, J.D.P.-R., G.Q., and J.K.; Formal Analysis, J.K., and J.D.P.-R.; Investigation, J.D.P.-R., G.Q., V.K., J.P., O.P.-A., and P.R.; Resources, M.V., G.Q., and J.K.; Data Curation, G.Q., J.D.P.-R.; Writing—Original Draft Preparation, M.V., and J.K.; Writing—Review & Editing, J.D.P.-R.; Visualization, J.D.P.-R., and J.K.; Supervision, S.A., M.V., and J.K.; Project Administration, S.A., M.V., and J.K.; Funding Acquisition, S.A., M.V., and J.K. All authors have read and agreed to the published version of the manuscript.

Funding: This research was funded by Instituto de Salud Carlos III [CP16/00034, PI17/00127, and PI20/00964], the RETICS funded by the PN 2018–2021 (Spain), ISCIII-Sub-Directorate General for Research Assessment and Promotion and the European Regional Development Fund (FEDER) [RD16/0022]; and the Foundation for Pediatric Research in Finland and Special Governmental Subsidy for Clinical Research.

Institutional Review Board Statement: The study was conducted according to the guidelines of the Declaration of Helsinki and approved by the Ethics Committee of Helsinki University Hospital (HUS §59/2015). The ethical approval number of the project is 123/13/03/03/2015.

Informed Consent Statement: Informed consent was obtained from all subjects involved in the study.

Data Availability Statement: Metabolomics data are available on Zenodo (https://zenodo.org/record/4495124#.YBpwoC1DkWp (2 February 2021)).

Acknowledgments: The authors would like to express their gratitude to the babies and their parents who enrolled their infants and gave their consent for this study.

Conflicts of Interest: The authors declare no conflict of interest.

References

1. Van der Linde, D.; Konings, E.E.M.; Slager, M.A.; Witsenburg, M.; Helbing, W.A.; Takkenberg, J.J.M.; Roos-Hesselink, J.W. Birth Prevalence of Congenital Heart Disease Worldwide: A Systematic Review and Meta-Analysis. *J. Am. Coll. Cardiol.* **2011**, *58*, 2241–2247. [CrossRef] [PubMed]
2. Bellinger, D.C.; Wypij, D.; Rivkin, M.J.; DeMaso, D.R.; Robertson, R.L.; Dunbar-Masterson, C.; Rappaport, L.A.; Wernovsky, G.; Jonas, R.A.; Newburger, J.W. Adolescents with D-Transposition of the Great Arteries Corrected with the Arterial Switch Procedure: Neuropsychological Assessment and Structural Brain Imaging. *Circulation* **2011**, *124*, 1361–1369. [CrossRef] [PubMed]
3. Lim, J.M.; Kingdom, T.; Saini, B.; Chau, V.; Post, M.; Blaser, S.; Macgowan, C.; Miller, S.P.; Seed, M. Cerebral Oxygen Delivery Is Reduced in Newborns with Congenital Heart Disease. *J. Thorac. Cardiovasc. Surg.* **2016**, *152*, 1095–1103. [CrossRef] [PubMed]
4. Sun, L.; Macgowan, C.K.; Sled, J.G.; Yoo, S.-J.; Manlhiot, C.; Porayette, P.; Grosse-Wortmann, L.; Jaeggi, E.; McCrindle, B.W.; Kingdom, J.; et al. Reduced Fetal Cerebral Oxygen Consumption Is Associated with Smaller Brain Size in Fetuses With Congenital Heart Disease. *Circulation* **2015**, *131*, 1313–1323. [CrossRef]
5. Park, I.S.; Yoon, S.Y.; Min, J.Y.; Kim, Y.H.; Ko, J.K.; Kim, K.S.; Seo, D.M.; Lee, J.H. Metabolic Alterations and Neurodevelopmental Outcome of Infants with Transposition of the Great Arteries. *Pediatr. Cardiol.* **2006**, *27*, 569–576. [CrossRef]
6. Van der Laan, M.E.; Verhagen, E.A.; Bos, A.F.; Berger, R.M.F.; Kooi, E.M.W. Effect of Balloon Atrial Septostomy on Cerebral Oxygenation in Neonates with Transposition of the Great Arteries. *Pediatr. Res.* **2013**, *73*, 62–67. [CrossRef]
7. Mukherjee, D.; Lindsay, M.; Zhang, Y.; Lardaro, T.; Osen, H.; Chang, D.C.; Brenner, J.I.; Abdullah, F. Analysis of 8681 Neonates with Transposition of the Great Arteries: Outcomes with and without Rashkind Balloon Atrial Septostomy. *Cardiol. Young* **2010**, *20*, 373–380. [CrossRef]
8. Hiremath, G.; Natarajan, G.; Math, D.; Aggarwal, S. Impact of Balloon Atrial Septostomy in Neonates with Transposition of Great Arteries. *J. Perinatol.* **2011**, *31*, 494–499. [CrossRef]
9. Hypoxic-Ischemic Encephalopathy: A Review for the Clinician | Cerebrovascular Disease | JAMA Pediatrics | JAMA Network. Available online: https://jamanetwork.com/journals/jamapediatrics/fullarticle/2118582 (accessed on 13 July 2020).
10. Terraneo, L.; Samaja, M. Comparative Response of Brain to Chronic Hypoxia and Hyperoxia. *Int. J. Mol. Sci* **2017**, *18*, 1914. [CrossRef]
11. Torres-Cuevas, I.; Parra-Llorca, A.; Sánchez-Illana, A.; Nuñez-Ramiro, A.; Kuligowski, J.; Cháfer-Pericás, C.; Cernada, M.; Escobar, J.; Vento, M. Oxygen and Oxidative Stress in the Perinatal Period. *Redox Biol.* **2017**, *12*, 674–681. [CrossRef]
12. Redox Signaling during Hypoxia in Mammalian Cells | Elsevier Enhanced Reader. Available online: https://reader.elsevier.com/reader/sd/pii/S2213231717302355?tken=9F2B800D98BB8EAC354ADDF5283F82835DF2689AA00C41C7A45BC70C1B1C24E5 567F1645B0354F153F5ECD84CE7FAF8E (accessed on 13 July 2020).
13. Piñeiro-Ramos, J.D.; Núñez-Ramiro, A.; Llorens-Salvador, R.; Parra-Llorca, A.; Sánchez-Illana, Á.; Quintás, G.; Boronat-González, N.; Martínez-Rodilla, J.; Kuligowski, J.; Vento, M.; et al. Metabolic Phenotypes of Hypoxic-Ischemic Encephalopathy with Normal vs. Pathologic Magnetic Resonance Imaging Outcomes. *Metabolites* **2020**, *10*, 109. [CrossRef]
14. Wu, T.-W.; Tamrazi, B.; Hsu, K.-H.; Ho, E.; Reitman, A.J.; Borzage, M.; Blüml, S.; Wisnowski, J.L. Cerebral Lactate Concentration in Neonatal Hypoxic-Ischemic Encephalopathy: In Relation to Time, Characteristic of Injury, and Serum Lactate Concentration. *Front. Neurol.* **2018**, *9*, 293. [CrossRef]
15. Fanos, V.; Pintus, R.; Dessì, A. Clinical Metabolomics in Neonatology: From Metabolites to Diseases. *NEO* **2018**, *113*, 406–413. [CrossRef] [PubMed]
16. Sánchez-Illana, Á.; Thayyil, S.; Montaldo, P.; Jenkins, D.; Quintás, G.; Oger, C.; Galano, J.-M.; Vigor, C.; Durand, T.; Vento, M.; et al. Novel Free-Radical Mediated Lipid Peroxidation Biomarkers in Newborn Plasma. *Anal. Chim. Acta* **2017**, *996*, 88–97. [CrossRef] [PubMed]
17. Sánchez-Illana, Á.; Shah, V.; Piñeiro-Ramos, J.D.; Di Fiore, J.M.; Quintás, G.; Raffay, T.M.; MacFarlane, P.M.; Martin, R.J.; Kuligowski, J. Adrenic Acid Non-Enzymatic Peroxidation Products in Biofluids of Moderate Preterm Infants. *Free Radic. Biol. Med.* **2019**, *142*, 107–112. [CrossRef]
18. Broadhurst, D.; Goodacre, R.; Reinke, S.N.; Kuligowski, J.; Wilson, I.D.; Lewis, M.R.; Dunn, W.B. Guidelines and Considerations for the Use of System Suitability and Quality Control Samples in Mass Spectrometry Assays Applied in Untargeted Clinical Metabolomic Studies. *Metabolomics* **2018**, *14*, 1–17. [CrossRef] [PubMed]
19. Kessner, D.; Chambers, M.; Burke, R.; Agus, D.; Mallick, P. ProteoWizard: Open Source Software for Rapid Proteomics Tools Development. *Bioinformatics* **2008**, *24*, 2534–2536. [CrossRef]

20. Smith, C.A.; Want, E.J.; O'Maille, G.; Abagyan, R.; Siuzdak, G. XCMS: Processing Mass Spectrometry Data for Metabolite Profiling Using Nonlinear Peak Alignment, Matching, and Identification. *Anal. Chem.* **2006**, *78*, 779–787. [CrossRef]
21. Benton, H.P.; Want, E.J.; Ebbels, T.M.D. Correction of Mass Calibration Gaps in Liquid Chromatography-Mass Spectrometry Metabolomics Data. *Bioinformatics* **2010**, *26*, 2488–2489. [CrossRef]
22. Tautenhahn, R.; Böttcher, C.; Neumann, S. Highly Sensitive Feature Detection for High Resolution LC/MS. *BMC Bioinform.* **2008**, *9*, 504. [CrossRef]
23. Kuhl, C.; Tautenhahn, R.; Böttcher, C.; Larson, T.R.; Neumann, S. CAMERA: An Integrated Strategy for Compound Spectra Extraction and Annotation of Liquid Chromatography/Mass Spectrometry Data Sets. *Anal. Chem.* **2012**, *84*, 283–289. [CrossRef]
24. Sánchez-Illana, Á.; Piñeiro-Ramos, J.D.; Sanjuan-Herráez, J.D.; Vento, M.; Quintás, G.; Kuligowski, J. Evaluation of Batch Effect Elimination Using Quality Control Replicates in LC-MS Metabolite Profiling. *Anal. Chim. Acta* **2018**, *1019*, 38–48. [CrossRef]
25. Quintás, G.; Sánchez-Illana, Á.; Piñeiro-Ramos, J.D.; Kuligowski, J. Chapter Six-Data Quality Assessment in Untargeted LC-MS Metabolomics. In *Comprehensive Analytical Chemistry*; Jaumot, J., Bedia, C., Tauler, R., Eds.; Data Analysis for Omic Sciences: Methods and Applications; Elsevier: Amsterdam, The Netherlands, 2018; Volume 82, pp. 137–164.
26. Kuligowski, J.; Sánchez-Illana, Á.; Sanjuán-Herráez, D.; Vento, M.; Quintás, G. Intra-Batch Effect Correction in Liquid Chromatography-Mass Spectrometry Using Quality Control Samples and Support Vector Regression (QC-SVRC). *Analyst* **2015**, *140*, 7810–7817. [CrossRef]
27. Chang, C.-C.; Lin, C.-J. LIBSVM: A Library for Support Vector Machines. *ACM Trans. Intell. Syst. Technol.* **2011**, *2*, 1–27. [CrossRef]
28. Chong, J.; Soufan, O.; Li, C.; Caraus, I.; Li, S.; Bourque, G.; Wishart, D.S.; Xia, J. MetaboAnalyst 4.0: Towards More Transparent and Integrative Metabolomics Analysis. *Nucleic Acids Res.* **2018**, *46*, W486–W494. [CrossRef]
29. Chouchani, E.T.; Pell, V.R.; Gaude, E.; Aksentijević, D.; Sundier, S.Y.; Robb, E.L.; Logan, A.; Nadtochiy, S.M.; Ord, E.N.J.; Smith, A.C.; et al. Ischaemic Accumulation of Succinate Controls Reperfusion Injury through Mitochondrial ROS. *Nature* **2014**, *515*, 431–435. [CrossRef]
30. Johnston, M.V.; Trescher, W.H.; Ishida, A.; Nakajima, W. Neurobiology of Hypoxic-Ischemic Injury in the Developing Brain. *Pediatr. Res.* **2001**, *49*, 735–741. [CrossRef]
31. Fatemi, A.; Wilson, M.A.; Johnston, M.V. Hypoxic-Ischemic Encephalopathy in the Term Infant. *Clin. Perinatol.* **2009**, *36*, 835–858. [CrossRef]
32. Teshima, Y.; Akao, M.; Li, R.A.; Chong, T.H.; Baumgartner, W.A.; Johnston, M.V.; Marbán, E. Mitochondrial ATP-Sensitive Potassium Channel Activation Protects Cerebellar Granule Neurons from Apoptosis Induced by Oxidative Stress. *Stroke* **2003**, *34*, 1796–1802. [CrossRef]
33. Millán, I.; Piñero-Ramos, J.D.; Lara, I.; Parra-Llorca, A.; Torres-Cuevas, I.; Vento, M. Oxidative Stress in the Newborn Period: Useful Biomarkers in the Clinical Setting. *Antioxidants* **2018**, *7*, 193. [CrossRef]
34. Stincone, A.; Prigione, A.; Cramer, T.; Wamelink, M.M.C.; Campbell, K.; Cheung, E.; Olin-Sandoval, V.; Grüning, N.-M.; Krüger, A.; Tauqeer Alam, M.; et al. The Return of Metabolism: Biochemistry and Physiology of the Pentose Phosphate Pathway. *Biol. Rev. Camb Philos Soc.* **2015**, *90*, 927–963. [CrossRef]
35. Brekke, E.M.; Morken, T.S.; Widerøe, M.; Håberg, A.K.; Brubakk, A.-M.; Sonnewald, U. The Pentose Phosphate Pathway and Pyruvate Carboxylation after Neonatal Hypoxic-Ischemic Brain Injury. *Biol. Rev.* **2014**, *34*, 724–734. [CrossRef]
36. Kuehne, A.; Emmert, H.; Soehle, J.; Winnefeld, M.; Fischer, F.; Wenck, H.; Gallinat, S.; Terstegen, L.; Lucius, R.; Hildebrand, J.; et al. Acute Activation of Oxidative Pentose Phosphate Pathway as First-Line Response to Oxidative Stress in Human Skin Cells. *Mol. Cell* **2015**, *59*, 359–371. [CrossRef]
37. Ralser, M.; Wamelink, M.M.; Kowald, A.; Gerisch, B.; Heeren, G.; Struys, E.A.; Klipp, E.; Jakobs, C.; Breitenbach, M.; Lehrach, H.; et al. Dynamic Rerouting of the Carbohydrate Flux Is Key to Counteracting Oxidative Stress. *J. Biol.* **2007**, *6*, 10. [CrossRef] [PubMed]
38. Wang, Y.-P.; Zhou, L.-S.; Zhao, Y.-Z.; Wang, S.-W.; Chen, L.-L.; Liu, L.-X.; Ling, Z.-Q.; Hu, F.-J.; Sun, Y.-P.; Zhang, J.-Y.; et al. Regulation of G6PD Acetylation by SIRT2 and KAT9 Modulates NADPH Homeostasis and Cell Survival during Oxidative Stress. *EMBO J.* **2014**, *33*, 1304–1320. [CrossRef]
39. Akram, M.; Shah, S.M.A.; Munir, N.; Daniyal, M.; Tahir, I.M.; Mahmood, Z.; Irshad, M.; Akhlaq, M.; Sultana, S.; Zainab, R. Hexose Monophosphate Shunt, the Role of Its Metabolites and Associated Disorders: A Review. *J. Cell. Physiol.* **2019**, *234*, 14473–14482. [CrossRef] [PubMed]
40. Vissers, M.C.M.; Das, A.B. Potential Mechanisms of Action for Vitamin C in Cancer: Reviewing the Evidence. *Front. Physiol.* **2018**, *9*, 809. [CrossRef] [PubMed]

*antioxidants*

MDPI

Article

The Mechanism of the Neuroprotective Effect of Kynurenic Acid in the Experimental Model of Neonatal Hypoxia–Ischemia: The Link to Oxidative Stress

Ewelina Bratek-Gerej [1,*], Apolonia Ziembowicz [1], Jakub Godlewski [2] and Elzbieta Salinska [1]

1   Department of Neurochemistry, Mossakowski Medical Research Institute, Polish Academy of Sciences, 02-106 Warsaw, Poland; aziembowicz@imdik.pan.pl (A.Z.); elasalin@gmail.com (E.S.)
2   Tumor Microenvironment Laboratory, Mossakowski Medical Research Institute, Polish Academy of Sciences, 02-106 Warsaw, Poland; jgodlewski@imdik.pan.pl
*   Correspondence: ebratek@imdik.pan.pl

Abstract: The over-activation of NMDA receptors and oxidative stress are important components of neonatal hypoxia–ischemia (HI). Kynurenic acid (KYNA) acts as an NMDA receptor antagonist and is known as a reactive oxygen species (ROS) scavenger, which makes it a potential therapeutic compound. This study aimed to establish the neuroprotective and antioxidant potential of KYNA in an experimental model of HI. HI on seven-day-old rats was used as an experimental model. The animals were injected i.p. with different doses of KYNA 1 h or 6 h after HI. The neuroprotective effect of KYNA was determined by the measurement of brain damage and elements of oxidative stress (ROS and glutathione (GSH) level, SOD, GPx, and catalase activity). KYNA applied 1 h after HI significantly reduced weight loss of the ischemic hemisphere, and prevented neuronal loss in the hippocampus and cortex. KYNA significantly reduced HI-increased ROS, GSH level, and antioxidant enzyme activity. Only the highest used concentration of KYNA showed neuroprotection when applied 6 h after HI. The presented results indicate induction of neuroprotection at the ROS formation stage. However, based on the presented data, it is not possible to pinpoint whether NMDA receptor inhibition or the scavenging abilities are the dominant KYNA-mediated neuroprotective mechanisms.

Keywords: neonatal hypoxia–ischemia; kynurenic acid (KYNA); oxidative stress; neuroprotection

Citation: Bratek-Gerej, E.; Ziembowicz, A.; Godlewski, J.; Salinska, E. The Mechanism of the Neuroprotective Effect of Kynurenic Acid in the Experimental Model of Neonatal Hypoxia–Ischemia: The Link to Oxidative Stress. *Antioxidants* **2021**, *10*, 1775. https://doi.org/10.3390/antiox10111775

Academic Editors: Julia Kuligowski and Máximo Vento

Received: 23 September 2021
Accepted: 2 November 2021
Published: 5 November 2021

Publisher's Note: MDPI stays neutral with regard to jurisdictional claims in published maps and institutional affiliations.

1. Introduction

Despite recent advances in neonatology, perinatal asphyxia (hypoxia–ischemia), remains a common problem in clinical practice, occurring in 2–4 per 1000 live births at term [1]. While acute asphyxia occurs in 1 per 1000 live births, mild hypoxia ensues more frequently, affecting up to 2% of all childbirths [2,3]. A hypoxic–ischemic (HI) insult is the result of a transient or permanent interruption of the blood and, thus, the oxygen supply to the infant's brain during delivery [4].

Severe asphyxia results in multi-organ failure and may lead to perinatal fatalities [5]. The most significant problem from a healthcare standpoint is children with a medium level of HI encephalopathy, which often leads to severe developmental disorders, including cerebral palsy, cognitive impairment, and/or epilepsy [6]. The standard-of-care treatment for asphyxiated neonates remains unsatisfactory, and, despite applied procedures, traces of neonatal asphyxia remain in the form of lifelong neurological and intellectual deficits. While some therapeutic avenues to prevent asphyxia-related complications in full-term neonates (e.g., hypothermia) exist, pharmacological approaches are still lacking [7]. Therefore, the development of effective and safe drug regimens that could be used in humans in various forms of brain ischemia, including birth asphyxia, is a subject of research efforts.

The accumulation of reactive oxygen species (ROS) is one of the most important factors involved in HI brain injury. An increased concentration of ROS shifts the antioxidant/oxidant balance towards the latter, thus initiating oxidative stress. Cells are equipped to combat oxidative stress and to neutralize ROS; superoxide dismutase (SOD), catalase (CAT), and glutathione peroxidase (GPx) supported by glutathione are well-known antioxidant enzymes [8]. However, in HI conditions, this antioxidant cell defense often fails.

HI induces the over-activation of many proteins, including hypoxia-induced factor-1α (HIF-1α). HIF-1α translocation to the nucleus stimulates pro-apoptotic genes, including the Bcl-2 family and apoptosis-inducing factor (AIF), but also the expression of sentinel proteins, such as poly (ADP-ribose) polymerase-1 (PARP-1), inducing further injury [9].

The immature brain displays high excitability, mostly due to a developmental increase in the expression of NMDA and α-amino-3-hydroxy-5-methyl-isoxazole-4-propionic acid (AMPA) receptors, and their different subunit composition when compared to the adult brain [10]. Moreover, the neonatal brain is vulnerable to oxidative stress damage due to lower levels of antioxidants. These differences account for differential responses to anti-ischemic insult treatments. A variety of inhibitors of NMDA receptors have been shown to have protective effects in adult hypoxic ischemia. However, it was found that the application of the NMDA receptor antagonist in neonates may cause abnormal neurodegeneration because activation of the NMDA receptor is required for the normal development of the brain [11,12]. Therefore, the safety and long-term effects of applying NMDA receptor antagonists for HIE treatment require a factor that will block receptors for a limited time.

Kynurenic acid (KYNA), an endogenous molecule produced in the kynurenine pathway of tryptophan metabolism, is known as an endogenous modulator of glutamatergic and cholinergic neurotransmission [13–15]. Its involvement in mechanisms initiated by ischemia was signalized by the information that the level of KYNA in the brain and cerebrospinal fluid increases significantly, but only for a short time after ischemia, and that its distribution matches the infarct brain regions [16,17].

KYNA is an endogenous antagonist of NMDA receptors binding to the glycine site [18], and this KYNA property was implied to exert neuroprotection in various preclinical models of hypoxic–ischemic brain injury [19–21]. However, the involvement of other KYNA properties, such as anti-inflammatory and receptor-independent anti-oxidative properties, in neuroprotection has not been studied [22,23].

KYNA, when administered peripherally, penetrates the blood–brain barrier, thus being a promising therapeutic agent to decrease excitotoxicity [24]. Another advantage of KYNA is that it cannot be metabolized to excitotoxic agents, and scavenges oxygen radicals, decreasing cellular damage [25]. The application of KYNA in high concentrations or for a prolonged time causes damage to neuronal cells [26,27]; however, KYNA applied in small doses is quickly removed from the organism and only accumulates in the liver [28].

This study aimed to establish the neuroprotective and antioxidant potential of KYNA administered in small doses in an experimental model of birth asphyxia.

2. Materials and Methods

2.1. Ethics Approval and Consent for Participation

All described experiments were conducted according to the guidelines of the Declaration of Helsinki, approved by the 4th Local Ethical Committee (263/2017) based in Warsaw, Poland, and were performed following Polish governmental regulations (Dz.U.97.111.724), the European Community Council Directive of 24 November 1986 (86/609/EEC) and Directive 2010/63/EU.

The animals came from breeding run by the Mossakowski Medical Research Institute's animal facility. Each experiment was performed on 3 different litters (10–12 rats per litter), and animals were randomly selected for experimental groups (2–3 animals from each litter). All surgeries were performed under anesthesia, and all efforts were made to minimize animal suffering and the number of animals used. The mortality rate did not exceed 5%.

2.2. Experimental Hypoxia–Ischemia on 7-Day-Old Rats

Hypoxia–ischemia (H-I) was induced according to Rice et al. [29] with small modifications [30]. Briefly, seven-day-old Wistar rat pups of both sexes were anesthetized with sevoflurane. The left common carotid artery was exposed and cut between double ligatures of silk sutures, or was only exposed (sham control). After 60 min of recovery, the pups were placed for 75 min in a humidified chamber filled with a hypoxic gas mixture (7.3% oxygen in nitrogen, 35 °C). After hypoxic treatment, the pups were returned to their cages and housed with their mothers.

2.3. Drug Application

KYNA (Tocris Bioscience, Bristol, UK) (300, 150, 50 mg/kg of body weight) was administered intraperitoneally at 1 h or 6 h after HI. The doses of KYNA were determined based on our previous experiments and the literature [24]. Sham-operated and HI control rats were injected with saline.

2.4. Evaluation of Brain Damage

Fourteen days after the insult (at PND21), the rats were anesthetized with a lethal dose of Morbital and decapitated. The cerebral hemispheres were weighed separately and brain damage was assessed by the deficit in weight of the ipsilateral (left) hemisphere, expressed as a percentage of the weight of the contralateral hemisphere (to the nearest 0.1 mg).

Histological assessment of brain damage was performed on brains isolated seven days after HI. Animals were anesthetized and then perfused transcardially with phosphate-buffered saline (PBS) followed by fixation solution (4% paraformaldehyde in PBS, pH 7.4). Brains were removed and postfixed for 3 h at 4 °C in the same fixing solution. Then brains were cryoprotected overnight in 30% sucrose solution, frozen on dry ice, and stored at −70 °C. Brains were cut into 20–30 μm coronal sections on a cryostat. Sections were stained with 0.5% cresyl violet according to the Nissl staining protocol for histological assessment of neuronal cell damage. The number of survived cells was counted under 200-fold magnification in the cortex in the visual field (250 μm × 250 μm) and the CA1 area of the hippocampus (100 μm in length) using AxioVision imaging software (Carl Zeiss, Aalen, Germany).

2.5. Tissue Preparation for Biochemical Analysis

Brain samples for biochemical analyses were collected 4 h after HI and 3 h after KYNA injections. The rats were decapitated, and the brain tissue samples containing the cerebral cortex and hippocampus were taken from both hemispheres for further examination. Tissues from the ipsilateral and contralateral hemispheres were homogenized separately in buffers appropriate for each analysis. The protein concentration of the homogenates was determined by the Bradford method and the homogenates were used for further analyses.

2.6. Determination of Oxidative Stress

2.6.1. Determination of ROS Level

The levels of ROS in brain hemispheres were measured using 2,7-dichlorofluorescein acetate (DCF-DA, Invitrogen Molecular Probes, Eugene, OR, USA). Brain homogenates were placed in 40 mM Tris-HCL buffer (pH 7.4) and incubated with 2.5 μM DCF-DA in a 96-well plate for 30 min at 37 °C. The DCF fluorescence was then detected with a multifunctional microplate reader (FLUOstar Omega, BMG LABTECH, Ortenberg, Germany) at 488 nm excitation and 530 nm emission wavelength. The relative fluorescence units (RFUs) of the homogenates were calculated per 1 mg of protein.

2.6.2. Determination of Glutathione Concentration

Brain tissue homogenates in 25 mM HEPES buffer (pH 7.4) containing 250 mM sucrose were centrifuged at $1000\times g$ for 5 min at 4 °C. The supernatants were collected to measure the glutathione concentration using the Fluorimetric Glutathione Assay Kit (Sigma-Aldrich, St. Louis, MO, USA) according to the manufacturer's instructions.

2.6.3. Determination of Antioxidant Enzyme Activity

Superoxide Dismutase

Brain tissue homogenates suspended in 20 mM HEPES buffer (pH 7.2), containing 1 mM EGTA, 210 mM mannitol, and 70 mM sucrose per 1 g of tissue, were centrifuged at $1500\times g$ for 5 min at 4 °C.

The supernatant was collected to determine SOD activity using the Superoxide Dismutase Assay Kit (Cayman Chemical, Ann Arbor, MI, USA) according to the instructions provided by the manufacturer. The activity of the enzyme is expressed as the number of enzymatic units per milligram of protein (U/mg protein).

Glutathione Peroxidase (GPx)

Brain tissue homogenates suspended in 50 mM Tris-HCl buffer (pH 7.5) containing 5 mM EDTA and 1 mM dithiothreitol (DTT) per 1 g of tissue were centrifuged at $10{,}000\times g$ for 15 min at 4 °C. The supernatants were collected to determine GPx activity using the Glutathione Peroxidase Assay Kit (Cayman Chemical, Ann Arbor, MI, USA) according to the instructions provided by the manufacturer.

Catalase

Homogenates suspended in 50 mM potassium orthophosphate buffer (pH 7.0) containing 1 mM EDTA were centrifuged at $10{,}000\times g$ for 15 min at 4 °C. The supernatants were collected for enzyme activity determination using the Catalase Assay Kit (Cayman Chemical, Ann Arbor, MI, USA) according to the instructions provided by the manufacturer.

2.7. Determination of HIF-1α Concentration

Brain tissue homogenates suspended in PBS were centrifuged at $5000\times g$ for 5 min and assayed using a Rat HIF-1α ELISA Kit (MyBioSource Inc., San Diego, CA, USA) according to the user manual.

2.8. Statistical Analysis

The results are expressed as the mean $\pm$ SEM of each experimental group. Statistical analysis was performed using a one-way ANOVA test with Dunnett's post hoc test for significant differences between groups (GraphPad Prism 5; GraphPad Software Inc., La Jolla, CA, USA). The differences were considered statistically significant when the *p*-value was less than 0.05.

3. Results

3.1. The Effect of KYNA Application on HI-Induced Brain Damage

Initially, we assessed the protective effect of KYNA on HI-induced brain damage. The HI-evoked 42% weight deficit of the ipsilateral hemisphere was significantly reduced by the application of KYNA in a dose-dependent manner 1 h after HI ($p < 0.01$) (Figure 1). A similar reduction in the weight deficit was also observed when KYNA was applied 6 h after HI, although the protective effect was less pronounced. These results highlighted the neuroprotective potential of KYNA.

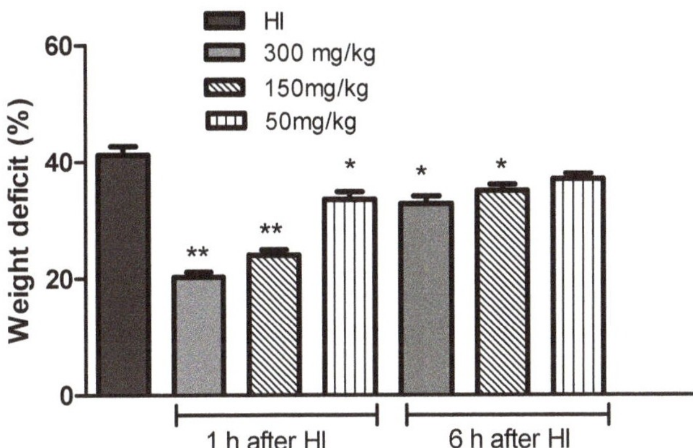

Figure 1. Effect of the application of KYNA on the weight deficit of the ipsilateral hemisphere after HI. KYNA (at all tested doses) was applied i.p. 1 h or 6 h after HI insult. The weight deficit is expressed as the percentage of the weight of the contralateral (right) hemisphere. The results are presented as the mean $\pm$ SEM, n = 4–9; * $p < 0.05$, ** $p < 0.01$ compared to the HI group.

Histological analysis upon HI treatment showed, expectedly, marked cell loss in the cortex, and damage and disorganization of neurons in the CA1 region of the hippocampus in the HI group (Figure 2A–C). The number of surviving neurons observed in the central part of the CA1 region was reduced by 55% (Figure 2B), and by 75.6% in the cortex (Figure 2C), compared to the control.

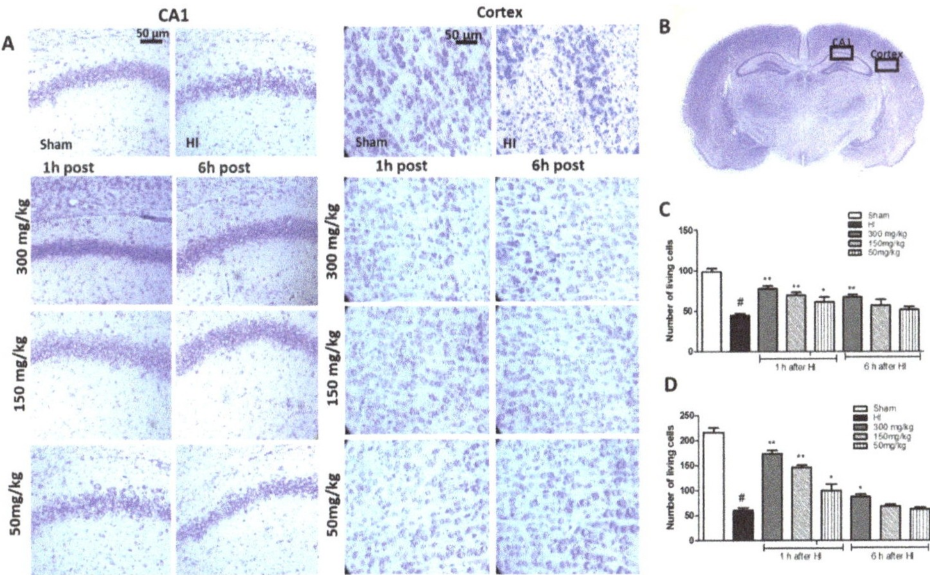

Figure 2. The effect of KYNA application 1 h or 6 h after HI on cell survival was observed in the CA1 region of the hippocampus and the cerebral cortex of the ipsilateral hemisphere 7 days after HI. (**A**) The microphotographs show the ipsilateral hemisphere. Scale bar represents 50 µm. (**B**) Localization of analyzed brain regions. (**C**) Quantification of surviving neurons in the central part of the CA1 region and (**D**) cortex. The results are presented as the mean $\pm$ SEM, n = 6; statistically significant differences: * $p < 0.05$, ** $p < 0.01$ compared to the HI group; # $p < 0.001$ compared to the sham-operated group.

A detailed histological analysis of KYNA-instigated changes in the brain, which could shed more light on the KYNA neuroprotective effect, revealed that KYNA applied 1 h after HI largely prevented changes in the CA1 region of the hippocampus, and significantly decreased neuronal loss in the cortex. KYNA in a dose of 300 mg/kg, applied 6 h after HI, increased the number of surviving neurons in the CA1 region and in the cortex to 68% and 40% of the control, respectively. However, KYNA applied in lower doses did not prevent the loss of neurons in either the CA1 region of the hippocampus or in the cortex.

The application of KYNA to sham-operated animals did not affect the weight of the brain hemispheres, and HI did not change the weight of the contralateral hemisphere (data not shown).

The results presented above clearly indicated a strong KYNA-mediated neuroprotective effect resulting from treatment 1 h after HI, and this effect was still observed when KYNA was applied 6 h later. These results defined the therapeutic window for KYNA, showing the limitations of the use of low doses.

Due to the weak neuroprotective effect of lower doses of KYNA applied 6 h after HI, these doses were not included in further analyses.

3.2. The Effect of KYNA Application on Changes in ROS Level in Rat Brain after HI

We then assessed the extent of KYNA-mediated alterations in the overall ROS levels to pinpoint the mechanism of KYNA-induced neuroprotection. HI increased the levels of ROS in the left ischemic hemisphere to more than 250% of the control, while remaining stable in the right hemisphere (Figure 3). KYNA treatment significantly prevented the rise in ROS levels in a dose-dependent manner, suggesting, for the first time, the mechanism beyond KYNA-mediated neuroprotection in hypoxia–ischemia.

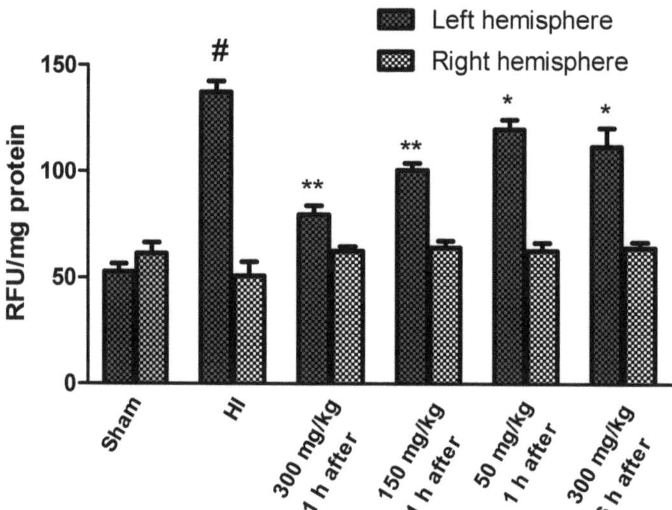

Figure 3. Effects of KYNA application on changes in ROS levels observed in the rat brain after HI. The results are presented as the means ± SEM, $n = 6$; statistically significant differences: * $p < 0.05$, ** $p < 0.001$ compared to the HI group; # $p < 0.001$ compared to the sham-operated group.

3.3. The Effect of KYNA Application on HI-Induced Changes in Antioxidant Enzymes Activity

HI increased the SOD activity to 454% of the control in the left (ipsilateral) ischemic hemisphere (Figure 4). KYNA applied 1 h after HI significantly reduced SOD activity in a dose-dependent manner. KYNA injection 6 h after HI in a dose of 300 mg/kg body weight resulted in a significant decrease in SOD activity to 333% of the control. However, the application of KYNA in doses of 50 and 150 mg/kg body weight did not result in a statistically significant decrease in SOD activity (data not shown).

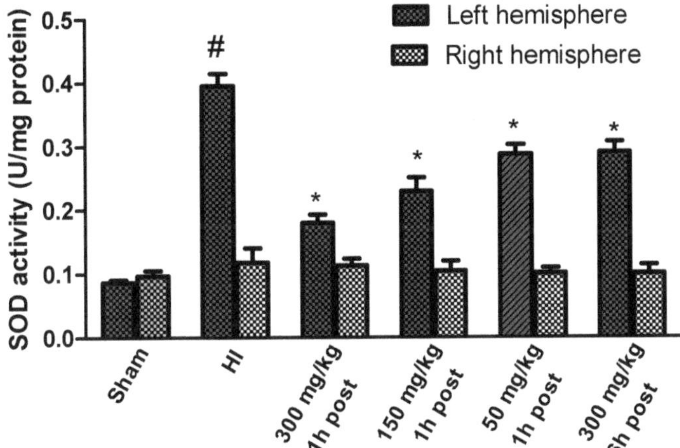

Figure 4. Effect of KYNA application on HI-induced changes in SOD activity. The results are presented as the means ± SEM, $n = 6$; statistically significant differences: * $p < 0.001$ compared to the HI group; # $p < 0.001$ compared to the sham-operated group.

HI, in this experimental model, did not change the activity of SOD in the contralateral hemisphere compared to the sham-operated group.

HI resulted in a significant increase in GPx activity in the left hemisphere to 440% of the control level, while the enzyme activity in the right hemisphere remained unchanged (Figure 5). KYNA in a dose of 300 mg/kg and 150 mg/kg, applied 1 after HI, significantly decreased the GPx activity in the left hemisphere to 290% and 354% of the control, respectively. KYNA in a dose of 50 mg/kg also reduced the activity of GPx (to 380% of the control), although this decrease was not statistically significant. KYNA injected 6 h after HI did not result in a decrease in GPx activity.

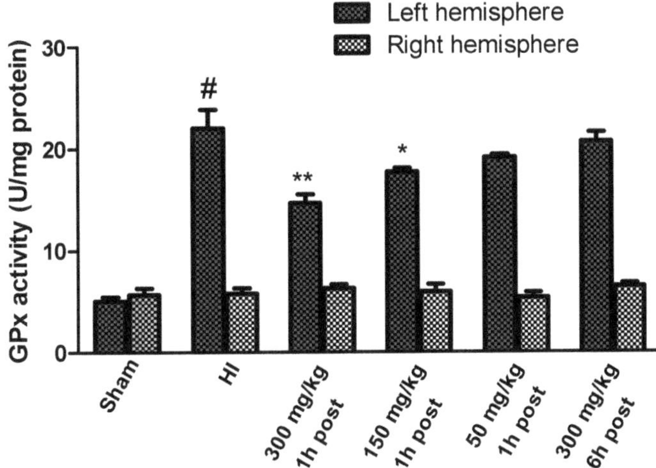

Figure 5. Effect of KYNA application on HI-induced changes in glutathione peroxidase activity. The results are presented as the means ± SEM, $n = 6–7$; statistically significant differences: * $p < 0.05$, ** $p < 0.01$ compared to the HI group; # $p < 0.001$ compared to the sham-operated group.

The GSH concentration determined in brains isolated from control rats ranged between 29.7 and 30.25 nmol/mg of protein in the left and right hemispheres, respectively (Figure 6).

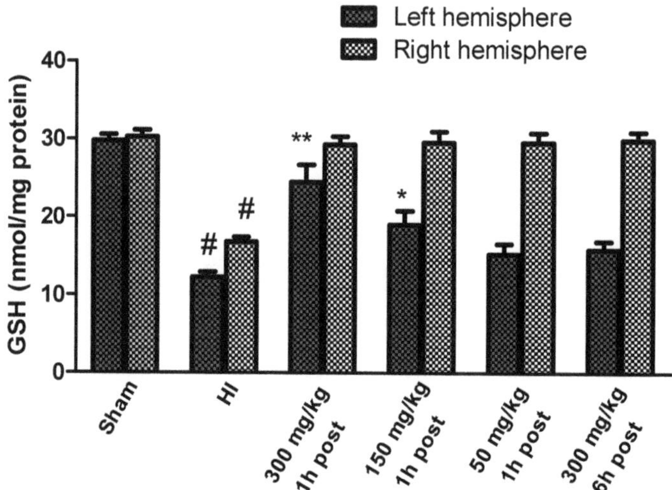

Figure 6. Effect of KYNA application on changes in the GSH concentration in the brains of rat pups after HI. The results are presented as the means $\pm$ SEM, n = 5–6; statistically significant differences: * $p < 0.01$, ** $p < 0.001$ compared to the HI group, # $p < 0.001$ compared to the sham-operated group.

HI resulted in a significant decrease in GSH in both hemispheres to 41.3% and 55.5% of the control in the left and right hemispheres, respectively (Figure 6). KYNA applied in doses of 300 mg/kg and 150 mg/kg body weight 1 h after HI significantly restored the GSH concentration, whereas the application of KYNA in a dose of 50 mg/kg body weight did not result in a significant increase in GSH concentration compared to the HI group, similarly to 300 mg/kg applied 6 h after HI. However, KYNA in each of the investigated doses restored GSH concentration in the right hemisphere to the control level, independently of the time of application.

The activity of the catalase increased significantly after HI only in the left hemisphere, reaching a value of 269% of the control (Figure 7). KYNA applied 1 h after HI resulted in a significant dose-dependent decrease in catalase activity. KYNA injected 6 h after HI, in a dose of 300 mg/kg body weight, also resulted in a decrease in catalase activity to 210% of the control.

3.4. The Effect of KYNA Application on the Changes in HIF-1α Concentration Observed after HI

The HIF-1α concentration measured in the brains of the sham-operated rats ranged from 11.85 to 10.97 ng/mg protein in the left and right hemispheres, respectively. HI significantly increased the HIF-1α concentration to 26.6 ng/mg protein in the left hemisphere, and to 18.65 ng/mg protein in the right hemisphere, which is 225% and 169.5% of the control, respectively (Figure 8). The application of KYNA in a dose of 300 mg/kg and 150 mg/kg body weight 1 h after HI significantly decreased the HIF-1α concentration to 168% and 177% of the control, respectively.

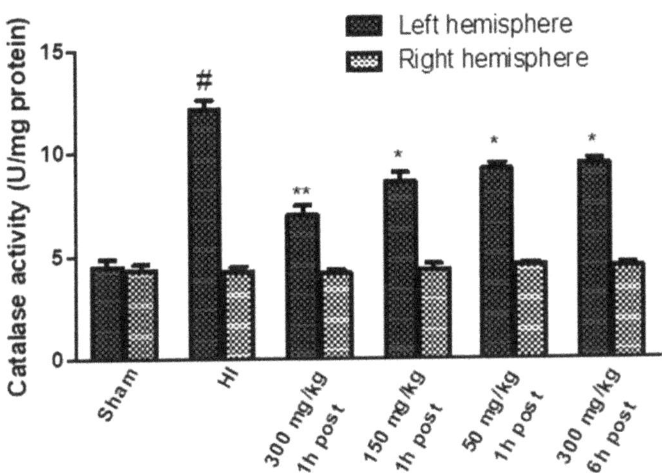

Figure 7. Effect of KYNA application on HI-induced changes in catalase activity. The results are presented as the means $\pm$ SEM, n = 6; statistically significant differences: * $p < 0.005$, ** $p < 0.001$ compared to the HI group; # $p < 0.001$ compared to the sham-operated group.

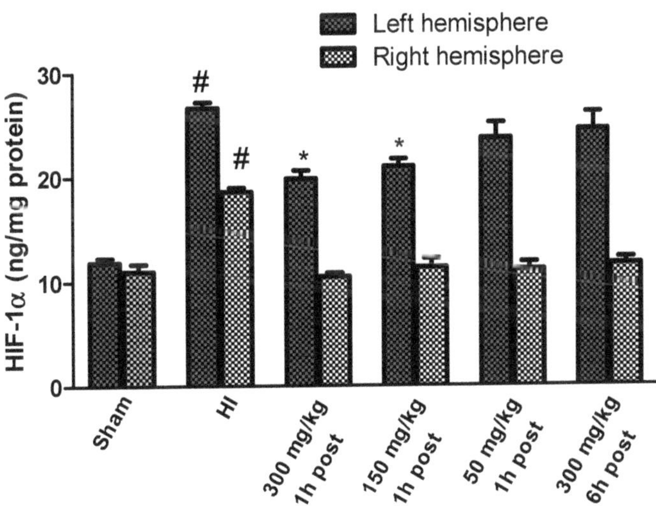

Figure 8. Effect of KYNA application on HI-induced changes in the HIF-1α concentration. The results are presented as the means $\pm$ SEM, n = 6; statistically significant differences: * $p < 0.001$, compared to the HI group; # $p < 0.001$ compared to the sham-operated group.

However, KYNA in a dose of 50 mg/kg body weight did not result in a statistically significant decrease in the HIF-1α concentration, and the application of KYNA in a dose of 300 mg/kg body weight 6 h after HI did not reduce the HIF-1α concentration in the left, ischemic hemisphere; however, KYNA application at all the used doses significantly reduced the HI-increased concentration of HIF-1α in the right hemisphere, restoring it to the control level.

4. Discussion

The results of the present study demonstrate a protective effect of the early application of KYNA on the development of neuronal injury in a rat model of perinatal asphyxia.

Our results show that the administration of KYNA in a dose of 300 mg/kg of body weight results in neuroprotection if it takes place up to 6 h after HI. KYNA in smaller doses (50 and 150 mg/kg) is only effective when applied immediately after HI. The application of KYNA 1 h after HI significantly reduced weight loss in the ischemic hemisphere, and reduced neuronal loss in the CA1 region of the hippocampus and cortex, whereas the neuroprotective effect of KYNA applied 6 h after HI was negligible.

It was shown that after profound asphyxia, a "latent" phase develops, typically lasting approximately 6 h, during which the brain can still recover from the insult, only to die hours to days later after a "secondary" deterioration, characterized by seizures, cytotoxic edema, and progressive failure of cerebral oxidative metabolism. Therefore, accepted treatments and experimental therapies of HI should be initiated before the onset of secondary deterioration [31–33]. Our results indicate a short therapeutic window for KYNA treatment, which is in line with the generally accepted treatment initiation time. The observed neuroprotective effects of KYNA, expressed as a reduction in neuronal loss and brain damage, agree with previous observations [21,34]; however, KYNA application temporal boundaries have been demonstrated for the first time.

Excessive glutamate release and excitotoxic NMDA receptor activation are important mechanisms of neuronal damage during primary energy loss and the reoxygenation/reperfusion phases of HIE development. Studies have shown that pretreatment with the NMDA receptor antagonist MK-801 provided only a partially effective operation in a piglet model of HIE [35], and MK-801 had not only protective, but also toxic, effects in rat pups [36]. Therefore, the use of alternative agents inhibiting the over-excitation of NMDA receptors is of great interest.

KYNA's property to inhibit NMDA receptors is mainly associated with its neuroprotective action, but its effective antioxidant properties and hydroxyl radical scavenging capacity may also play a role in neuroprotection [22,37]. HI generates oxidative stress manifested by the increased generation of ROS. It was shown that key antioxidant enzymes increase their activity after HI, although the level of GSH decreases, probably as an effect of intensive consumption in a reaction catalyzed by GPx [31,38]. KYNA was shown to reduce ROS levels and regulate antioxidant enzyme activity in vivo in an experimental model of oxidative stress induced by an injection of quinolinic acid into a rat striatum, and in vitro on rat brain samples and *Xenopus laevis* oocytes by inducing the Fenton reaction [22,39].

Our results show, for the first time, that the application of KYNA 1 h after HI significantly decreases ROS levels and antioxidant enzyme activity. We also observed partial restoration of the GSH concentration. However, the application of KYNA 6 h after HI had a much weaker effect, again suggesting that there is only a short therapeutic window for KYNA.

Low oxygen activates hypoxia-inducible factor (HIF) transcription factors that play a dominant role in coordinating the transcriptional response to hypoxia. HIF-1α regulates a multitude of genes involved in glycolysis, inflammation, apoptosis, and proteolysis. The functional HIF-1 complex is formed by regulatory subunit-α (HIF-1α) and a constitutively expressed β-subunit [40]. Under normoxia, HIF-1α is rapidly degraded; however, in hypoxic conditions, its accumulation may lead to the activation of genes, such as *Nox2*, that encode the pro-oxidant enzyme NADPH oxidase, which is a major source of cellular ROS [41].

The presented results show that the application of KYNA 1 h after HI reduced the HIF-1α protein levels that were increased by HI. It is difficult to determine whether this decrease in HIF-1α is the result of a reduction in ROS production or an inhibition of NMDA receptors. ROS act as an important signal molecule on MAPK, PI3K/Act/mTOR, and NF-κB pathways, which regulate the expression of HIF-1α [42]. In hypoxic conditions, increased ROS activates NF-κB, which plays a key role in the transcription of HIF-1α.

Moreover, it was shown that reactive nitrogen species (RNS) formed by endogenous ROS and NO inactivate HIF prolyl 4-hydroxylases (PHDs) that degrade HIF-1α [42]. On the other hand, the inhibition of NMDA receptors by MK-801 was shown to inhibit HIF-1α expression, suggesting that this mechanism may also be involved in the operation of KYNA [43]. The lack of an effect of KYNA applied 6 h after HI on the HIF-1α level again indicates the short therapeutic window for KYNA.

5. Conclusions

KYNA-mediated neuroprotection observed in the HI model of birth asphyxia is, in large, connected with the reduction in oxidative stress.

KYNA reduces ROS production, and the presented results indicate that this reduction is not the result of the antioxidant enzymes' mobilization. This suggests that the induction of neuroprotection at the ROS formation stage could be the result of KYNA's ability to inhibit NMDA receptors and prevent calcium-induced mechanisms, leading to mitochondrial damage. However, it can also be the result of the scavenging abilities of KYNA and the direct reduction in produced ROS. Unfortunately, based on the presented data, it is not possible to pinpoint the dominant KYNA-mediated neuroprotective mechanisms, and further investigations are required. The therapeutic effect of relatively small doses of KYNA only manifests when the application is performed a short time after HI, and this therapeutic window for KYNA fits in a commonly accepted time of intervention. The presented results demonstrate KYNA's potential as a promising new therapeutic agent.

Author Contributions: Conceptualization, E.B.-G. and E.S.; methodology, E.B.-G. and A.Z.; investigation, E.B.-G. and A.Z.; writing—original draft preparation, E.B.-G. and E.S.; writing—review and editing, J.G.; supervision, E.S.; funding acquisition, J.G. and E.B.-G. All authors have read and agreed to the published version of the manuscript.

Funding: This research was partially funded by the National Science Center of Poland (2020/39/B/NZ5/02893) (to J.G.).

Institutional Review Board Statement: All described experiments were conducted according to the guidelines of the Declaration of Helsinki, approved by the 4th Local Ethical Committee (263/2017) based in Warsaw, Poland, and were performed following Polish governmental regulations (Dz.U.97.111.724), the European Community Council Directive of 24 November 1986 (86/609/EEC) and Directive 2010/63/EU.

Informed Consent Statement: Not applicable.

Data Availability Statement: The data presented in this study are available in this manuscript.

Conflicts of Interest: The authors declare no conflict of interest.

References

1. Placha, K.; Luptakova, D.; Baciak, L.; Ujhazy, E.; Juranek, I. Neonatal brain injury as a consequence of insufficient cerebral oxygenation. *Neuro Endocrinol. Lett.* **2016**, *37*, 79–96. [PubMed]
2. Bryce, J.; Boschi-Pinto, C.; Shibuya, K.; Black, R. WHO estimates of the causes of death in children. *Lancet* **2005**, *365*, 1147–1152. [CrossRef]
3. Rochala, K.; Krasomski, G. The estimation of foetus' incorrect biophysical profile influence on Intraventricular Hemorrhage (IVH) and asphyxiation prevalence in infant. *Nowa Pediatr.* **2013**, *2*, 44–52.
4. Mazumder, M.K.; Paul, R.; Bhattacharya, P.; Borah, A. Neurological sequel of chronic kidney disease: From diminished Acetylcholinesterase activity to mitochondrial dysfunctions, oxidative stress and inflammation in mice brain. *Sci. Rep.* **2019**, *9*, 3097. [CrossRef]
5. Greco, P.; Nencini, G.; Piva, I.; Scioscia, M.; Volta, C.A.; Spadaro, S.; Neri, M.; Bonaccorsi, G.; Greco, F.; Cocco, I.; et al. Pathophysiology of hypoxic–ischemic encephalopathy: A review of the past and a view on the future. *Acta Neurol. Belg.* **2020**, *120*, 277–288. [CrossRef]
6. Łosiowski, Z. *Uszkodzenia Układu Nerwowego Związane z Patologią Okresu Płodowego i Okołoporodowego. Neurologia Dziecięca*; Państwowy Zakład Wydawnictw Lekarskich: Warszawa, Poland, 1999; pp. 326–332.
7. Davidson, J.O.; Wassink, G.; van den Heuij, L.G.; Bennet, L.; Gunn, A.J. Therapeutic hypothermia for neonatal hypoxic-ischemic encephalopathy-where to from here? *Front. Neurol.* **2015**, *6*, 198. [CrossRef] [PubMed]

8. Taysi, S.; Tascan, A.S.; Ugur, M.G.; Demir, M.; Uuro, M.G. Radicals, oxidative/nitrosative stress and preeclampsia. *Mini Rev. Med. Chem.* **2019**, *19*, 178–193. [CrossRef]

9. Herrera-Marschitz, M.; Perez-Lobos, R.; Lespay-Rebolledo, C.; Tapia-Bustos, A.; Casanova-Ortiz, E.; Morales, P.; Valdes, J.L.; Bustamante, D.; Cassels, B.K. Targeting sentinel proteins and extrasynaptic glutamate receptors: A therapeutic strategy for preventing the effects elicited by Perinatal Asphyxia? *Neurotox. Res.* **2018**, *33*, 461–473. [CrossRef]

10. Gennaro, M.; Mattiello, A.; Pizzorusso, T. Rodent models of developmental ischemic stroke for translational research: Strengths and weaknesses. *Neural Plast.* **2019**, *2019*, 5089321. [CrossRef] [PubMed]

11. Olney, J.W.; Wozniak, D.F.; Jevtovic-Todorovic, V.; Farber, N.B.; Bittigau, P.; Ikonomidou, C. Drug-induced apoptotic neurodegeneration in the developing brain. *Brain Pathol.* **2002**, *12*, 488–498. [CrossRef]

12. Adesnik, H.; Li, G.; During, M.J.; Pleasure, S.J.; Nicoll, R.A. NMDA receptors inhibit synapse unsilencing during brain development. *Proc. Natl. Acad. Sci. USA* **2008**, *105*, 5597–5602. [CrossRef] [PubMed]

13. Fiedorowicz, M.; Choragiewicz, T.; Thaler, S.; Schuettauf, F.; Nowakowska, D.; Wojtunik, K.; Reibaldi, M.; Avitabile, T.; Kocki, T.; Turski, W.A.; et al. Tryptophan and kynurenine pathway metabolites in animal models of retinal and optic nerve damage: Different dynamics of changes. *Front. Physiol.* **2019**, *10*, 1254. [CrossRef] [PubMed]

14. Kemp, J.A.; Foster, A.C.; Leeson, P.D.; Priestley, T.; Tridgett, R.; Iversen, L.L.; Woodruff, G.N. 7-Chlorokynurenic acid is a selective antagonist at the glycine modulatory site of the N-methyl-D-aspartate receptor complex. *Proc. Natl. Acad. Sci. USA* **1988**, *85*, 6547–6550. [CrossRef] [PubMed]

15. Kessler, M.; Terramani, T.; Lynch, G.; Baudry, M. A glycine site associated with N-methyl-D-aspartic acid receptors: Characterization and identification of a new class of antagonists. *J. Neurochem.* **1989**, *52*, 1319–1328. [CrossRef]

16. Ceresoli-Borroni, G.; Schwarcz, R. Neonatal asphyxia in rats: Acute effects on cerebral kynurenine metabolism. *Pediatr. Res.* **2001**, *50*, 231–235. [CrossRef]

17. Mangas, A.; Yajeya, J.; González, N.; Ruiz, I.; Duleu, S.; Geffard, M.; Coveñas, R. Overexpression of kynurenic acid in stroke: An endogenous neuroprotector? *Ann. Anat. Anat. Anz.* **2017**, *211*, 33–38. [CrossRef]

18. Hilmas, C.; Pereira, E.F.; Alkondon, M.; Rassoulpour, A.; Schwarcz, R.; Albuquerque, E.X. The brain metabolite kynurenic acid inhibits alpha7 nicotinic receptor activity and increases non-alpha7 nicotinic receptor expression: Physiopathological implica-tions. *J. Neurosci.* **2001**, *21*, 7463–7473. [CrossRef] [PubMed]

19. Colpo, G.D.; Venna, V.R.; McCullough, L.D.; Teixeira, A.L. Systematic review on the involvement of the Kynurenine Pathway in stroke: Pre-clinical and clinical evidence. *Front. Neurol.* **2019**, *10*, 778. [CrossRef]

20. Hertelendy, P.; Toldi, J.; Fülöp, F.; Vécsei, L. Ischemic stroke and kynurenines: Medicinal chemistry aspects. *Curr. Med. Chem.* **2019**, *25*, 5945–5957. [CrossRef]

21. Nozaki, K.; Beal, M.F. Neuroprotective effects of L-Kynurenine on Hypoxia—Ischemia and NMDA lesions in neonatal rats. *Br. J. Pharmacol.* **1992**, *12*, 400–407. [CrossRef]

22. Lugo-Huitrón, R.; Blanco-Ayala, T.; Ugalde-Muñiz, P.; Carrillo-Mora, P.; Pedraza-Chaverrí, J.; Adaya, I.D.S.; Maldonado, P.D.; Torres, I.; Pinzón, E.; Ortiz-Islas, E.; et al. On the antioxidant properties of kynurenic acid: Free radical scavenging activity and inhibition of oxidative stress. *Neurotoxicol. Teratol.* **2011**, *33*, 538–547. [CrossRef] [PubMed]

23. Walczak, K.; Turski, W.A.; Rajtar, G. Kynurenic acid inhibits colon cancer proliferation in vitro: Effects on signaling pathways. *Amino Acids* **2014**, *46*, 2393–2401. [CrossRef] [PubMed]

24. Toldi, J.; Oláh, G.; Herédi, J.; Menyhárt, Á.; Czinege, Z.; Nagy, D.; Fuzik, J.; Krucsó, E.; Kocsis, K.; Knapp, L.; et al. Unexpected effects of peripherally administered kynurenic acid on cortical spreading depression and related blood–brain barrier permeability. *Drug Des. Dev. Ther.* **2013**, *7*, 981–987. [CrossRef]

25. Esquivel, D.G.; Ramírez-Ortega, D.; Pineda, B.; Castro, N.; Ríos, C.; de la Cruz, P.V. Kynurenine pathway metabolites and enzymes involved in redox reactions. *Neuropharmacology* **2017**, *112*, 331–345. [CrossRef]

26. Dabrowski, W.; Kwiecien, J.M.; Rola, R.; Klapec, M.; Stanisz, G.; Kotlinska-Hasiec, E.; Oakden, W.; Janik, R.; Coote, M.; Frey, B.N.; et al. Prolonged subdural infusion of kynurenic acid is associated with dose-dependent myelin damage in the rat spinal cord. *PLoS ONE* **2015**, *10*, e0142598. [CrossRef]

27. Langner, E.; Lemieszek, M.K.; Kwiecień, J.M.; Rajtar, G.; Rzeski, W.; Turski, W.A. Kynurenic acid induces impairment of oligodendrocyte viability: On the role of glutamatergic mechanisms. *Neurochem. Res.* **2017**, *42*, 838–845. [CrossRef] [PubMed]

28. Turska, M.; Pelak, J.; Turski, M.P.; Kocki, T.; Dukowski, P.; Plech, T.; Turski, W. Fate and distribution of kynurenic acid administered as beverage. *Pharm. Rep.* **2018**, *70*, 1089–1096. [CrossRef]

29. Rice, J.E., III; Vannucci, R.C.; Brierley, J.B. The influence of immaturity on hypoxic-ischemic brain damage in the rat. *Ann. Neurol.* **1981**, *9*, 131–141. [CrossRef] [PubMed]

30. Gerej, E.B.; Bronisz, A.; Ziembowicz, A.; Salinska, E. Pretreatment with mGluR2 or mGluR3 agonists reduces apoptosis induced by Hypoxia-Ischemia in Neonatal rat brains. *Oxidative Med. Cell. Longev.* **2021**, *2021*, 8848015. [CrossRef]

31. Bratek, E.; Ziembowicz, A.; Bronisz, A.; Salinska, E. The activation of group II metabotropic glutamate receptors protects neonatal rat brains from oxidative stress injury after hypoxia-ischemia. *PLoS ONE* **2018**, *13*, e0200933. [CrossRef] [PubMed]

32. Gunn, A.J.; Thoresen, M. Neonatal encephalopathy and hypoxic–ischemic encephalopathy. *Handb. Clin. Neurol.* **2019**, *162*, 217–237. [CrossRef] [PubMed]

33. Wintermark, P. Current controversies in newer therapies to treat birth asphyxia. *Int. J. Pediatr.* **2011**, *2011*, 1–5. [CrossRef]

34. Andiné, P.; Lehmann, A.; Ellrén, K.; Wennberg, E.; Kjellmer, I.; Nielsen, T.; Hagberg, H. The excitatory amino acid antagonist kynurenic acid administered after hypoxic-ischemia in neonatal rats offers neuroprotection. *Neurosci. Lett.* **1988**, *90*, 208–212. [CrossRef]

35. Yang, Z.-J.; Ni, X.; Carter, E.L.; Kibler, K.; Martin, L.J.; Koehler, R.C. Neuroprotective effect of acid-sensing ion channel inhibitor psalmotoxin-1 after hypoxia–ischemia in newborn piglet striatum. *Neurobiol. Dis.* **2011**, *43*, 446–454. [CrossRef]

36. Hattori, H.; Morin, A.M.; Schwartz, P.H.; Fujikawa, D.G.; Wasterlain, C.G. Posthypoxic treatment with MK-801 reduces hypoxic-ischemic damage in the neonatal rat. *Neurology* **1989**, *39*, 713. [CrossRef]

37. Tiszlavicz, Z.; Németh, B.; Fülöp, F.; Vécsei, L.; Tápai, K.; Ocsovszky, I.; Mándi, Y. Different inhibitory effects of kynurenic acid and a novel kynurenic acid analogue on tumour necrosis factor-α (TNF-α) production by mononuclear cells, HMGB1 pro-duction by monocytes and HNP1-3 secretion by neutrophils. *Naunyn Schmiedebergs Arch. Pharmacol.* **2011**, *383*, 447–455. [CrossRef]

38. Gamdzyk, M.; Ziembowicz, A.; Bratek, E.; Salinska, E. Combining hypobaric hypoxia or hyperbaric oxygen postconditioning with memantine reduces neuroprotection in 7-day-old rat hypoxia-ischemia. *Pharmacol. Rep.* **2016**, *68*, 1076–1083. [CrossRef]

39. Ferreira, F.; Schmitz, F.; Marques, E.; Siebert, C.; Wyse, A. Intrastriatal quinolinic acid administration impairs redox homeostasis and induces inflammatory changes: Prevention by kynurenic acid. *Neurotox. Res.* **2020**, *38*, 50–58. [CrossRef]

40. Semenza, G.L. HIF-1: Mediator of physiological and pathophysiological responses to hypoxia. *J. Appl. Physiol.* **2000**, *88*, 1474–1480. [CrossRef]

41. Yuan, G.; Khan, S.A.; Luo, W.; Nanduri, J.; Semenza, G.L.; Prabhakar, N.R. Hypoxia-inducible factor 1 mediates increased expression of NADPH oxidase-2 in response to intermittent hypoxia. *J. Cell. Physiol.* **2011**, *226*, 2925–2933. [CrossRef]

42. Shao, Y.; Wang, K.; Xiong, X.; Liu, H.; Zhou, J.; Zou, L.; Qi, M.; Liu, G.; Huang, R.; Tan, Z.; et al. The landscape of interactions between hypoxia-inducible factors and reactive oxygen species in the gastrointestinal tract. *Oxidative Med. Cell. Longev.* **2021**, *2021*, 8893663. [CrossRef]

43. Chen, X.; Wu, Q.; You, L.; Chen, S.; Zhu, M.; Miao, C. Propofol attenuates pancreatic cancer malignant potential via inhibition of NMDA receptor. *Eur. J. Pharmacol.* **2017**, *795*, 150–159. [CrossRef]

Article

Effects of Hypothermia and Allopurinol on Oxidative Status in a Rat Model of Hypoxic Ischemic Encephalopathy

Cristina Durán Fernández-Feijóo [1], Javier Rodríguez-Fanjul [2], Miriam Lopez-Abat [3], Stephanie Hadley [4], Mónica Cavia-Saiz [5], Pilar Muñiz [5], Juan Arnaez [6], José Ramón Fernández-Lorenzo [1] and Marta Camprubí Camprubí [3,*]

[1]   Department of Neonatology, Hospital Álvaro Cunqueiro, EOXI, 36312 Vigo, Spain; cris_dff@hotmail.com (C.D.F.-F.); joseramon.fernandez.lorenzo@usc.es (J.R.F.-L.)
[2]   Neonatal Intensive Care Unit, Paediatrics Department, Hospital Germans Trias i Pujol, Universitat Autonoma de Barcelona, 08916 Badalona, Spain; javier.rodriguez.fanjul@gmail.com
[3]   Department of Neonatology, BCNatal | Barcelona Center for Maternal Fetal and Neonatal Medicine Hospital Sant Joan de Déu and Hospital Clínic, University of Barcelona, 08950 Esplugues de Llobregat, Spain; mlopeza@fsjd.org
[4]   Vanderbilt University School of Medicine, Nashville, TN 37232, USA; stephanie.hadley@childrens.harvar.edu
[5]   Department of Biotechnology and Food Science, Facutlty of Sciences, University of Burgos, 09001 Burgos, Spain; monicacs@ubu.es (M.C.-S.); pminuiz@ubu.es (P.M.)
[6]   Department of Neonatology, Hospital Universitario de Burgos, NeNe Foundation, 09006 Burgos, Spain; jusoru@hotmail.com
*   Correspondence: mcamprubic@sjdhospitalbarcelona.org

Citation: Durán Fernández-Feijóo, C.; Rodríguez-Fanjul, J.; Lopez-Abat, M.; Hadley, S.; Cavia-Saiz, M.; Muñiz, P.; Arnaez, J.; Fernández-Lorenzo, J.R.; Camprubí Camprubí, M. Effects of Hypothermia and Allopurinol on Oxidative Status in a Rat Model of Hypoxic Ischemic Encephalopathy. *Antioxidants* **2021**, *10*, 1523. https://doi.org/10.3390/antiox10101523

Academic Editors: Julia Kuligowski and Máximo Vento

Received: 8 September 2021
Accepted: 20 September 2021
Published: 25 September 2021

Publisher's Note: MDPI stays neutral with regard to jurisdictional claims in published maps and institutional affiliations.

Abstract: Hypoxic ischemic encephalopathy (HIE) is one of the main causes of morbidity and mortality during the neonatal period, despite treatment with hypothermia. There is evidence that oxidative damage plays an important role in the pathophysiology of hypoxic-ischemic (HI) brain injury. Our aim was to investigate whether postnatal allopurinol administration in combination with hypothermia would reduce oxidative stress (OS) biomarkers in an animal model of HIE. Postnatal 10-day rat pups underwent unilateral HI of moderate severity. Pups were randomized into: Sham operated, hypoxic-ischemic (HI), HI + allopurinol (HIA), HI + hypothermia (HIH), and HI + hypothermia + allopurinol (HIHA). Biomarkers of OS and antioxidants were evaluated: GSH/GSSG ratio and carbonyl groups were tested in plasma. Total antioxidant capacity (TAC) was analyzed in plasma and cerebrospinal fluid, and 8-iso prostaglandin F2α was measured in brain tissue. Plasma 2,2′–azinobis-(3-ethyl-benzothiazoline-6-sulfonic acid) (ABTS) levels were preserved in those groups that received allopurinol and dual therapy. In cerebrospinal fluid, only the HIA group presented normal ferric reducing ability of plasma (FRAP) levels. Protein oxidation and lipid peroxidation were significantly reduced in all groups treated with hypothermia and allopurinol, thus enhancing neuroprotection in HIE.

Keywords: allopurinol; hypothermia; hypoxic-ischemic encephalopathy; oxidative stress; oxidative damage

1. Introduction

Hypoxic ischemic encephalopathy (HIE) is one of the main causes of mortality and long-term disability during the neonatal period [1,2]. Therapeutic hypothermia (TH) is now well established as standard treatment for infants with moderate-to-severe HIE. However, up to 25% of these infants die, and 20 to 60% survive with neurocognitive sequel [3].

Animal models allow for a better understanding of the pathophysiological mechanisms of hypoxic-ischemic brain injury (HI) and play a key role in investigating new therapeutic strategies [4].

HI triggers an increase in free radical production and subsequently oxidative stress (OS). Increased reactive oxygen and nitrogen species alter the function and/or structure of proteins, nucleic acids, and membrane lipids. The newborn brain is particularly susceptible to HI and OS damage due to its immaturity as well as its low concentrations of antioxidant defenses, increased production of superoxide ($O_2^{\bullet-}$), and a high content of free iron and polyunsaturated fatty acids [5].

There is evidence that oxidative damage plays an important role in the pathophysiology of HI brain injury [6,7]. Martini et al. classified the mechanism of injury in neonatal encephalopathy in three phases [8]: oxygen deprivation in Phase I (0–6 h), mitochondrial dysfunction and increased oxidative stress (lipid, protein, nucleic acid peroxidation) in Phase II (6–72), and epigenetic changes induced by free radicals, inflammation, and decreased neurogenesis in Phase III (>72 h).

Several neuroprotective therapies have been employed to neutralize the excess of free radicals, including therapeutic hypothermia (TH), allopurinol, melatonin, erythropoietin (EPO), and nitric oxide synthase (NOS) inhibitors [7–9]. Studies of TH suggest that it may decrease free radical production [10,11]. Similarly, allopurinol, a xanthine oxidase inhibitor and a free radical scavenger, has shown neuroprotective effects against hypoxia-reperfusion brain injury [12–14]. To our knowledge, there are no animal studies investigating the combined use of hypothermia and allopurinol as a neuroprotective strategy.

The aim of this study was to investigate whether postnatal allopurinol administration in combination with hypothermia would reduce oxidative stress in an animal model of HIE.

2. Materials and Methods

This study was performed following the Guide for the Care and Use of Laboratory Animals of the National Institutes of Health. The protocol was approved by the Ethics Committee for Animal Experimentation of the University of Barcelona (permit number 6575), following European (2010/63(UE)) and Spanish (RD 53/2013) regulations for the care and use of laboratory animals.

All surgeries were performed under inhaled isoflurane (2%), and all efforts were made to minimize the animals' suffering and the quantity of animals used. Animals were euthanized prior to the end of the experiments by administration of intra-peritoneal thiopental.

2.1. Experimental Design

On postnatal day 10 (P10), Wistar pup rats (HARLAM, Netherlands) weighing 12–14 g were used in this study. After birth, animals were kept with their mothers in cages with 12 h light/dark cycles at a constant temperature of 22 ± 1 °C with free access to food and water.

Hypoxia-ischemia (HI) was induced using the Rice–Vannucci model [15]. Seventy-one P10 rat pups underwent left common carotid artery ligation, following the described protocol. After <1 h of recovery with their mother, animals were exposed to 90 min of hypoxia (8% atmospheric oxygen) at 36–36.5 °C rectal temperature, which resulted in HI moderate insult [16,17]. Seven animals died (9.5%) after the procedure prior to being randomized. A total of 64 P10 rat pups were randomized into five experimental groups: sham-operated control (C), HI + normothermia (HI), HI + allopurinol (HIA), HI + hypothermia (HIH), and HI + hypothermia + allopurinol (HIHA).

Sham-operated animals were anesthetized, and a skin incision was performed to expose the left common carotid artery without artery ligation or hypoxia.

2.1.1. Allopurinol

All treatment groups received a single intraperitoneal injection of either allopurinol (zylosprim sodium, Burroughs Wellcome, Research Triangle Park, NC, USA) at 135 mg/kg (volume: 0.01 mL/g; HIA and HIHA groups) or saline (HI and HIH groups) depending on randomization, 15 min after the hypoxia procedure (before beginning the hypothermia or normothermia protocol), as described by Rodriguez-Fanjul et al. [18].

2.1.2. Hypothermia vs. Normothermia

At the end of the HI procedure, pups were also randomized into two groups: (i) those treated with systemic hypothermia (32.5–33 °C) and (ii) those maintained in normothermia (36–36.5 °C). All the pups were maintained in temperature-controlled chambers for 5 h and separated from each other to avoid rewarming. Temperature was continuously measured in one pup in each chamber with a rectal temperature probe (IT-21; Physitemp Instruments). After the treatment period, pups were immediately removed from the chamber and returned to their litter.

A diagram of the experimental design, including the number of animals used, is presented in Figure 1.

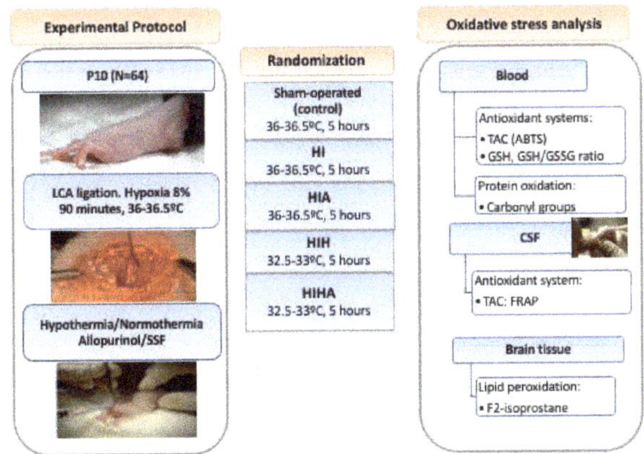

Figure 1. Study Diagram: Experimental Protocol, Randomization, and Oxidative stress analysis. P10: Ten days of life, LCA ligation: Left common carotid ligation. HI: Hypoxia-ischemia, HIA: Hypoxia-ischemia allopurinol, HIH: Hypoxia-ischemia hypothermia, HIHA: Hypoxia-ischemia hypothermia allopurinol. SSF: Physiologic serum.

2.2. Samples Obtention and Preparation

Seventy-two hours after the procedure, blood and cerebrospinal fluid (CSF) samples were obtained, processed, and stored at −80 °C until analysis. Pups were sacrificed immediately after sample collection according to protocol, and the brain was removed and stored at −80 °C.

Blood samples: Blood samples were collected in heparin tubes. Plasma was separated by centrifugation ($3000\times g$ rpm × 10 min) and stored at −80 °C. Hemolyzed blood plus distilled water was treated with cold chloroform: ethanol (3:5 v/v) and centrifuged ($3000\times g$ rpm × 10 min), and the supernatant was used for analysis.

CSF samples: CSF collection was performed as previously described [19] and stored at −80 °C.

Brain tissue samples: Coronal sections of 3 mm were cut under dry ice, and the hippocampus and cortico-subcortical area were located and dissected under microscopic visualization (Nikon SMZ645 (16–100 x)). Hippocampus and cortico-subcortical zones were extracted from the brain tissue, homogenized (10% w/w) in phosphate-buffered saline (PBS) with a pH of 7.4, and centrifuged ($2500\times g$ rpm × 10 min). The supernatant was then isolated and stored until analysis.

2.3. Sample Analysis

2.3.1. Plasma Total Antioxidant Capacity (TAC)

Total antioxidant capacity evaluates the overall antioxidant status of plasma and CSF as an estimation of the plasma capacity to neutralize oxidants. This technique is an overall measure of cumulative antioxidant status of biological fluids instead of individual antioxidants.

The total antioxidant capacity was evaluated in plasma and CSF. The ferric reducing ability power (FRAP) method was used in CSF. Due to the size of the pups, the volume of plasma was small, and some hemolysis was observed. As a result, the 2,2′-azino-bis-3-etilbenzothiazol-6-sulfonic acid (ABTS) method, unaffected by hemolysis, was used to evaluate TAC in plasma.

FRAP assay: This method was used to measure the presence of reducing agents in CSF [20]. This assay measures wavelength absorbance at 593 nm caused by the formation of blue-colored tripyridyl-s-triazine complexes (TPTZ) with ferric (II) [TPTZ-Fe (II)] in the presence of a reducing agent. The results are expressed as molar (mM) equivalents of Fe (II) (mM Fe (II)E). Fe (II) sulfate heptahydrate ($FeSO_4$ $7H_2O$) was purchased from Probus S.A. (Badalona, Spain).

ABTS method: This method was used to evaluate the ability of plasma antioxidants to scavenge the ABTS$^+$ radical, which absorbs wavelengths at 734 nm. 2,2′-Azino-bis(3-ethylbenzothiazoline-6-sulfonic acid) diammonium salt >98 (ABTS) was performed following the protocol outlined by Re et al. [21] and modified by Gonzalez et al. [22]. Briefly, the ABTS·+ radical cation is generated by the reaction of a 7 mM solution of ABTS in water with 2.45 mM potassium persulphate ($K_2S_2O_8$) >99% (1:1). The assay consists of 960 μL of ABTS$^+$, 35 μL of the buffer PBS 7 mM with a pH of 7.4, and 5 μL of the plasma sample. The inhibition of the absorbance of the ABTS$^+$ cation at 734 nm by the sample is measured after a 5 min incubation period. The results are expressed as molar equivalents (mM) of 6-hydroxy-2,5,7,8-tetramethylchroman-2-carboxylic acid 97% (TROLOX).

2.3.2. Plasma Glutathione Reduced/Oxidized (GSH/GSSG) Ratio Analysis

Plasma reduced glutathione (GSH) and oxidized glutathione (GSSG) levels were determined using the reaction between the sulfhydryl group of the GSH and 5,5′-dithio-bis-2-nitrobenzoic acid (DTNB, Ellman's reagent) [23]. The GSTNB (mixture between GSH and TNB) is reduced by glutathione reductase to recycle GSH and produce more TNB. The levels of TNB produced were directly proportional to the concentration of GSH in the sample. Briefly, the extract of hemolyzed blood was mixed with DTNB, NADPH, and glutathione reductase (Sigma-Aldrich, St. Louis, MO, USA). The level of total GSH (reduced and oxidized) was evaluated by measuring absorbance at 412 nm at 2 min intervals for 20 min. Glutathione disulfide (GSSG) was measured using the same method after derivatizing the samples with 2-vinilpiridine, and GSH was estimated by subtracting GSSG from total GSH. The results are expressed as the GSH/GSSG ratio.

2.3.3. Plasma Protein Carbonyl Groups (CG)

Plasma protein oxidation was assessed with an estimation of carbonil groups formed using the protocol described by Levine et al. [24] which is based on the reaction of the carbonyl group with 2,4-dinitrophenylhydrazine (DNFH) under acidic conditions. Plasma samples were mixed with DNFH and incubated 1 h. After that, they were precipitated with 500 μL of 20% (*w/v*) of trichloroacetic acid, washed three times with ethanol/ethyl acetate (1:1 *v/v*), and centrifuged at 6000× *g* for 3 min to remove any free 2,4-DNPH. Finally, 1 mL of 6 M guanidine, pH 2.3, was added at the samples and were incubated in a 37 °C water bath for 30 min. Carbonyl groups were calculated by absorption spectrophotometry at 373 nm, using a molar absorption coefficient of 22,000 M^{-1} cm^{-1}. The carbonyl groups levels were normalized by the protein concentration in plasma, and the results are expressed as nmol/mg protein. Total protein concentration was determined by the Lowry method.

2.3.4. Brain 8-Iso-Prostaglandin F2α (8-iso-PGF2α) Quantification

Levels of 8-iso-PGF2α were measured in the hippocampus and cortico-subcortical area 72 h after the HI event. They were quantified by ELISA according to kit instructions (OxiSelect 8-iso-Prostaglandin F2α ELISA Kit). To be prepared for the analysis, samples were treated with NaOH at 45 °C for 2 h. In addition, 100 μL of concentrated (10N) HCl per 500 μL of hydrolyzed sample was added. After that, samples were centrifuged for 5 min at 12,000× g rpm.

2.4. Statistical Analysis

Results are expressed in mean and interquartile range. The Kruskal–Wallis test was used to detect differences between groups. Post hoc analysis was performed to evaluate inter-group differences. STATA v13 was utilized for statistical analysis.

3. Results

A total of 71 rat pups were used for this experiment. Seven (9.5%) died after the procedure. None of the animals treated with HIH or HIHA died. All the samples were obtained 72 h after the procedure. The mean animal weight was 19.2 ± 3.7 g. From each animal a maximum of 0.5–0.7 mL of total blood was obtained, and this was immediately processed to one of the experiments, with the goal of having enough representative samples of each biomarker. Due to the small blood sample, not all of the animals were tested for all the biomarkers.

3.1. Allopurinol Administration, Alone or in Combination with Hypothermia, Protects Total Antioxidant Capacity (TAC) after an HI Event

Plasma TAC levels measured by ABTS were lower in the HI group compared to HIA and HIHA ($p < 0.001$), but there were no differences found between HI and those animals treated only with hypothermia (HIH) ($p = 0.194$). Mean plasma TAC values are presented in Table S1 (Supplementary Materials). In the post hoc analysis, differences were also seen between HIHA and HIH groups ($p < 0.001$) (Figure 2A).

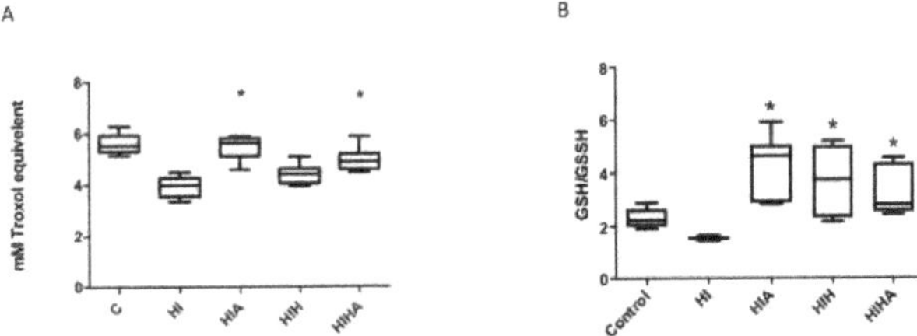

Figure 2. Antioxidant systems: (**A**) Box plot plasma ABTS levels; (**B**) Box plot plasma GSH/GSSG ratio. C: Control, HI: Hypoxia-ischemia, HIA: Hypoxia-ischemia + allopurinol, HIH: Hypoxia-ischemia + hypothermia, HIHA: Hypoxia-ischemia + hypothermia + allopurinol. * $p < 0.05$ in post-hoc analysis compared to HI.

3.2. GSH/GSSG Ratio Was Decreased after an HI Event, and Only Treated Groups Recovered to Normal Values

The plasma GSH/GSSG ratio was decreased in HI animals compared to those groups that received any treatment (HIA, HIH, HIHA) ($p < 0.001$). Levels in the HIA group were even higher than those in the C group ($p < 0.001$), as shown in Table S2 (Supplementary Materials) (Figure 2B).

3.3. TAC Levels in CSF Are Preserved When Allopurinol Is Administrated after an HI Event, but Not When Allopurinol Is Administrated in Combination with Hypothermia

HI animals showed the lowest levels of TAC in the CSF, as quantified by the FRAP method ($p = 0.011$). The HIA group presented similar levels to the C group ($p = 0.834$). No increases in TAC levels were detected in hypothermia-treated groups (HIH, HIHA). Results are presented in Table S3 (Supplementary Materials) (Figure 3).

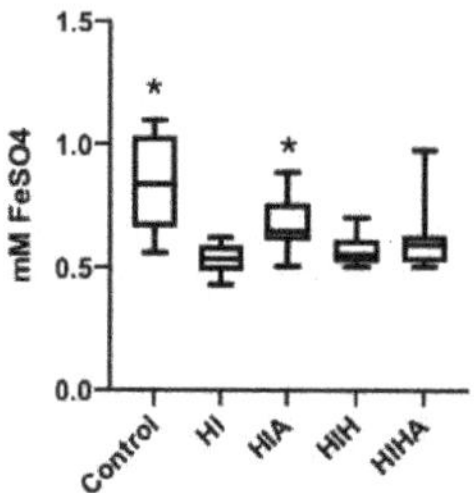

Figure 3. Box plot cerebral spinal fluid FRAP levels. C: Control, HI: Hypoxia-ischemia, HIA: Hypoxia-ischemia + allopurinol, HIH: Hypoxia-ischemia + hypothermia, HIHA: Hypoxia-ischemia + hypothermia + allopurinol. * $p < 0.05$ in post-hoc analysis compared to HI.

3.4. Protein Oxidation Increase after an HI Event, All Treatments Seem to Prevent Protein Oxidation

Plasma carbonyl group levels were significantly increased in the HI group. In the post hoc analysis, treated groups (HIA, HIH, HIHA) presented with lower plasma carbonyl group levels compared to HI ($p < 0.01$). Results are presented in Table S4 (Supplementary Materials) (Figure 4).

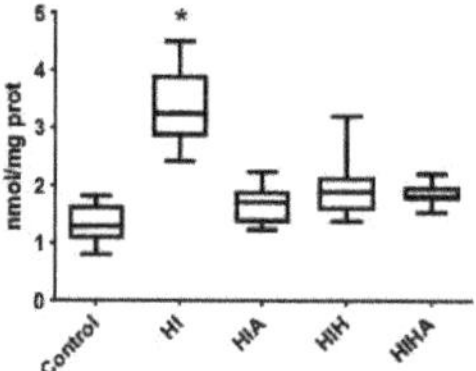

Figure 4. Box plot plasma carbonyl groups. C: Control, HI: Hypoxia-ischemia, HIA: Hypoxia-ischemia + allopurinol, HIH: Hypoxia-ischemia + hypothermia, HIHA: Hypoxia-ischemia + hypothermia + allopurinol. * $p < 0.05$ in post-hoc analysis compared to HI.

3.5. Lipid Peroxidation Increased after an HI Event and Decreased in the Treated Groups

Hippocampal 8-iso-PGF2α levels varied among groups ($p = 0.030$) (Figure 5a). In the cortical-subcortical area, 8-iso-PGF2α levels increased in the HI group ($p = 0.002$) and significantly decreased in HIH ($p < 0.001$) and HIHA ($p = 0.013$) (Figure 5b). Table S5 (Supplementary Materials).

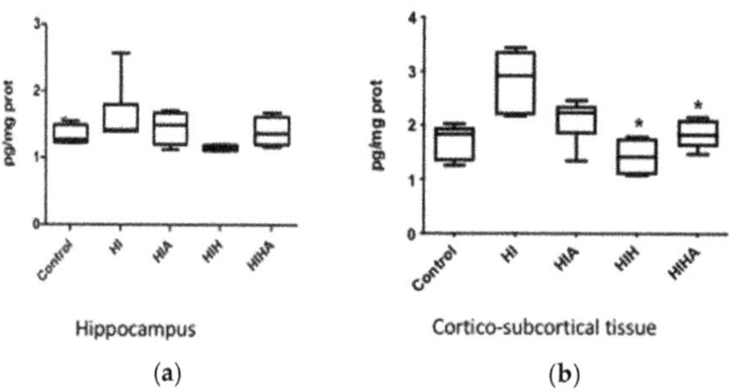

Figure 5. (a) Box plot of F2-isoprostane levels in hippocampus tissue. (b) Box plot of F2-isoprostane levels in cortico-subcortical tissue. C: Control, HI: Hypoxia-ischemia, HIA: Hypoxia-ischemia + allopurinol, HIH: Hypoxia-ischemia + hypothermia, HIHA: Hypoxia-ischemia + hypothermia + allopurinol. * $p < 0.05$ in post-hoc analysis compared to HI.

4. Discussion

There is increasing evidence supporting the role of oxidative stress in the pathogenesis of brain damage in HIE. Many studies have examined antioxidant therapies for treatment and have demonstrated their neuroprotective effect [18,25–27] despite their mechanism of action being unclear. In our previous paper, hypothermia and allopurinol were proven to provide a neuroprotective effect, improving histological, biochemical, and functional markers in a neonatal animal model of HIE [18]. In the same line, there is one ongoing clinical trial that is also studying the effect of allopurinol in addition to hypothermia in neonates with HI brain injury on neurocognitive outcome [28]. Despite these studies, nothing has been published about its effects on antioxidant systems and free radical production when hypothermia and allopurinol are administered together as a combined neuroprotective strategy.

After an HI event, a decrease in serum TAC levels was observed. These results are in agreement with the findings of other publications that demonstrate a reduction in TAC levels in cerebral tissue following HI injury [29,30]. In our study, TAC levels were preserved after the administration of the neuroprotective therapy with allopurinol, or the combination therapy of hypothermia and allopurinol. Curiously, Van Bel and colleagues, in their clinical trial using allopurinol in HIE patients, could not demonstrate this protective effect of allopurinol in relation to TAC levels. They hypothesized that this could be the result of the delayed administration of allopurinol (170 min after the insult), highlighting the importance of early administration of the antioxidant therapies after an HI insult [14]. Experimental studies have demonstrated that while free radical production begins with ischemia, it significantly increases during the reperfusion phase [31]. Ono and colleagues demonstrated that early administration of allopurinol decreased superoxide radical generation [32]. Another important point that supports early administration of allopurinol is related to the results observed in CSF TAC levels. Allopurinol is able to cross the blood brain barrier (BBB) [33]. Of note, while HIE increases BBB permeability [34], hypothermia causes its stabilization [35]. Analyzing CSF in our population, only the allopurinol group (HIA) maintained TAC levels similar to the control group, perhaps attributable to this BBB stabilization. These results support the perception that allopurinol must be administered as soon as possible before hypothermia, leading to BBB stabilization and, subsequently, difficult passage of allopurinol into the CNS. In this regard, Kaandorp JJ et al. presented the ALLO trial, in which allopurinol was administered antenatally when hypoxia was suspected [36]. In this study, a decrease in a biomarker of brain damage (S100B protein) was reported in the allopurinol group. Recently, based on the importance of the early allopurinol administration after an HI event, there is an ongoing multicenter trial (Phase III) studying the neuroprotection effect of allopurinol in addition to hypothermia treatment (ALBINO trial) in newborns with hypoxic-ischemic encephalopathy. In this study, allopurinol was administered intravenously within 30 min after birth to optimize the timing and inhibition of superoxide formation in asphyxiated infants with evolving HIE [28].

GSH is a critical non-enzymatic antioxidant protecting cells from OS. In our population, the GSH/GSSG ratio was decreased in the HI group. These data are in accordance with other published results [35], supporting the hypothesis that after an HI event there is a massive liberation of free radicals that consume reduced glutathione. Of note, groups undergoing neuroprotective treatments showed similar GSH/GSSG ratios as the control group. Moreover, it has been demonstrated that allopurinol plays a role in GSH normalization levels in experimental models of ischemia/reperfusion, renal failure, and respiratory distress [37–39]. In our study, GSH/GSSG ratios in the HIA group were remarkably higher, even higher than those of controls. This result opens a path to explore whether or not allopurinol, similar to other antioxidants such as melatonin [25], may have a role in the induction of antioxidant substances. Regarding hypothermia treatment, we observed a protective effect on the GSH/GSSG ratio consistent with previously described experimental models of cardiac circulatory arrest or cerebral hypoxia where hypothermia was induced [40,41].

In the same line, neuroprotection in terms of histology, biomarkers, and function were also demonstrated in our previous paper [18], where all the treatment groups presented an improvement in all of these aspects after treatment administration.

Protein oxidation and lipid peroxidation were also evaluated in our HIE model. Protein oxidation in HIE, measured by serum carbonyl group levels, has been previously proven in clinical and experimental essays [42–44], supporting our findings in the HI group. In our study, administration of any neuroprotective treatment (HIA, HIH, HIHA) seemed to prevent protein oxidation. Experimental studies have demonstrated that hypothermia decreases the degree of protein oxidation 3 to 6 h after an HI event [42,45]. It is not well known how allopurinol protects against protein oxidation, but this effect has also been described in heart disease studies [46].

Increased lipid peroxidation has also been reported after an HI event. There are many biomarkers used to evaluate this status. For example, 8-iso-PGF2α, a widely used OS biomarker [47], was analyzed in the hippocampus and cortico-subcortical zones. The elevated levels of 8-iso-PGF2α in the HI group are in accordance with other publications, demonstrating increases in lipid peroxidation products in the cortex and hippocampus in a neonatal rat model of HIE [48]. Lower levels of 8-iso-PGF2α were observed in all treated groups. Allopurinol effects on lipid peroxidation are not well established and, to our knowledge, have not been studied in HI animal models. However, it appeared to have a protective role in clinical cardiovascular trials, similar to its protection against protein oxidation [49].

There is also controversy regarding the effects of hypothermia on lipid peroxidation. Recently, Huun and colleagues demonstrated in a porcine animal model of HIE a decrease in urinary levels of 8-iso-PGF2α in those animals that were treated with hypothermia, but none of the other lipid peroxidation compounds evaluated were affected by the hypothermia treatment [50]. This group also analyzed lipid peroxidation in the cortex, subcortical white matter, and hippocampal tissue, finding a decrease in lipid peroxidation products only in subcortical white matter [51].

In our previous paper, Rodríguez-Fanjul J et al. [18], hypothermia, allopurinol + hypothermia, and allopurinol therapies were proven to confer neuroprotection after an HI event. Functional, histologic, and molecular improvement were described in all treated groups using the same protocol study. Histologically, damaged area and hippocampal volume were different among treatment groups. The largest tissue lesions were observed in the HI group, followed by HIA. From a molecular point of view, cleaved caspase 3 expression was increased in both HI and HIA. These results are in accordance with the decrease in the oxidative stress biomarkers that we observed in the HIHA and HIH groups. In the present study we also detected some positive effect against oxidative stress in the HIA group, reflecting the antioxidant effect of allopurinol. In fact, from a functional point of view, as shown in our previous paper [18], animals undergoing neuroprotection therapies, including the HIA group, presented an improvement in short-term (negative geotaxis) and long-term (Water Morris test) functional tests. Moreover, when the learning process was analyzed, no differences were found between treated groups. Animals from the HIA group had similar results to the HIH and HIHA groups.

Our paper reinforces the notion that the newborn brain is vulnerable to oxidative stress after an HI event. When considering all our data, it can be concluded that the administration of allopurinol, hypothermia, and the combination treatment (hypothermia + allopurinol) protects the brain against oxidative damage.

We observed changes in the expression of several oxidative stress biomarkers from different sources (serum, cerebral tissue, and CSF), and all of them support our previously histological biochemical and functional published results [18].

Finally, our data may also reinforce the importance of early administration of allopurinol treatment, even before initiating hypothermia, to achieve an optimal concentration in the CSF and improve and avoid an increased oxidative stress response. We believe that this finding may be imperative for future clinical trials of HIE that utilize allopurinol as a neuroprotective agent.

Supplementary Materials: The following are available online at https://www.mdpi.com/article/10.3390/antiox10101523/s1. Table S1. Plasma TAC levels measured using ABTS, Table S2. Plasma GSH/GSSG ratio, Table S3. Cerebral Spinal Fluid TAC levels, Table S4. Plasma carbonyl group levels, Table S5. Lipid peroxidation.

Author Contributions: M.C.C., J.R.F.-L. and C.D.F.-F.; methodology, J.R.-F., M.L.-A., M.C.-S., P.M. and J.A.; formal analysis, J.R.-F., M.L.-A., M.C.-S., P.M. and J.A.; investigation, C.D.F.-F. and J.R.-F.; resources M.C.C.; data curation, C.D.F.-F. and M.C.C.; writing—original draft preparation, C.D.F.-F.; writing—review and editing, S.H., J.R.F.-L and M.C.C.; supervision, M.C.C. All authors have read and agreed to the published version of the manuscript.

Funding: This research received no external funding.

Institutional Review Board Statement: This study was approved by the Ethics Committee for Animal Experimentation of the University of Barcelona (Permit Number 6575), following European (2010/63(UE)) and Spanish (RD 53/2013) regulations for the care and use of laboratory animals.

Informed Consent Statement: Not applicable.

Data Availability Statement: Data is contained within the article and Supplementary Materials.

Acknowledgments: The authors would like to acknowledge support from the Vanderbilt Medical Scholars Program and Isabel Salas for their technical support.

Conflicts of Interest: The authors declare no conflict of interest.

References

1. Blencowe, H.; Vos, T.; Lee, A.C.C.; Philips, R.; Lozano, R.; Alvarado, M.; Cousens, S.; Lawn, J.E. Estimates of neonatal morbidities and disabilities at regional and global levels for 2010: Introduction, methods overview, and relevant findings from the global burden of disease study. *Pediatr. Res.* **2013**, *74*, 4–16. [CrossRef] [PubMed]
2. Lee, A.C.C.; Kozuki, N.; Blencowe, H.; Vos, T.; Bahalim, A.; Darmstadt, G.L.; Niermeyer, S.; Ellis, M.; Robertson, N.J.; Cousens, S.; et al. Intrapartum-related neonatal encephalopathy incidence and impairment at regional and global levels for 2010 with trends from 1990. *Pediatr. Res.* **2013**, *74*, 50–72. [CrossRef] [PubMed]
3. Tagin, M.A.; Woolcott, C.G.; Vincer, M.J.; Whyte, R.K.; Stinson, D.A. Hypothermia for neonatal hypoxic ischemic encephalopathy. *Arch. Pediatr. Adolesc. Med.* **2012**, *166*, 558–566. [CrossRef] [PubMed]
4. Gunn, A.J.; Thoresen, M. Animal studies of neonatal hypothermic neuroprotection have translated well in to practice. *Resuscitation* **2015**, *97*, 88–90. [CrossRef]
5. Perrone, S.; Negro, S.; Tataranno, M.L.; Buonocore, G. Oxidative stress and antioxidant strategies in newborns. *J. Matern. Neonatal Med.* **2010**, *23*, 63–65. [CrossRef]
6. Blomgren, K.; Hagberg, H. Free radicals, mitochondria, and hypoxia–ischemia in the developing brain. *Free. Radic. Biol. Med.* **2006**, *40*, 388–397. [CrossRef]
7. Zhao, M.; Zhu, P.; Fujino, M.; Zhuang, J.; Guo, H.; Sheikh, I.; Zhao, L.; Li, X.-K. Oxidative stress in hypoxic-ischemic encephalopathy: Molecular mechanisms and therapeutic strategies. *Int. J. Mol. Sci.* **2016**, *17*, 2078. [CrossRef]
8. Martini, S.; Austin, T.; Aceti, A.; Faldella, G.; Corvaglia, L. Free radicals and neonatal encephalopathy: Mechanisms of injury, biomarkers, and antioxidant treatment perspectives. *Pediatr. Res.* **2019**, *87*, 823–833. [CrossRef]
9. Wang, Q.; Lv, H.; Lu, L.; Ren, P.; Li, L. Neonatal hypoxic–ischemic encephalopathy: Emerging therapeutic strategies based on pathophysiologic phases of the injury. *J. Matern. Neonatal Med.* **2018**, *32*, 3685–3692. [CrossRef]
10. Perrone, S.; Szabó, M.; Bellieni, C.V.; Longini, M.; Bangó, M.; Kelen, D.; Treszl, A.; Negro, S.; Tataranno, M.L.; Buonocore, G. Whole body hypothermia and oxidative stress in babies with hypoxic-ischemic brain injury. *Pediatr. Neurol.* **2010**, *43*, 236–240. [CrossRef] [PubMed]
11. Drury, P.P.; Gunn, E.R.; Bennet, L.; Gunn, A.J. Mechanisms of hypothermic neuroprotection. *Clin. Perinatol.* **2014**, *41*, 161–175. [CrossRef] [PubMed]
12. Palmer, C.; Vannucci, R.C.; Towfighi, J. Reduction of perinatal hypoxic-ischemic brain damage with allopurinol. *Pediatr. Res.* **1990**, *27*, 332–336. [CrossRef] [PubMed]
13. Shadid, M.; Moison, R.; Steendijk, P.; Hiltermann, L.; Berger, H.M.; van Bel, F. The effect of antioxidative combination therapy on post hypoxic-ischemic perfusion, metabolism, and electrical activity of the newborn brain. *Pediatr. Res.* **1998**, *44*, 119–124. [CrossRef] [PubMed]

14. Van Bel, F.; Shadid, M.; Moison, R.M.W.; Dorrepaal, C.A.; Fontijn, J.; Monteiro, L.; van de Bor, M.; Berger, H.M. Effect of allopurinol on postasphyxial free radical formation, cerebral hemodynamics, and electrical brain activity. *Pediatrics* **1998**, *101*, 185–193. [CrossRef]

15. Rice, J.E.; Vannucci, R.C.; Brierley, J.B. The influence of immaturity on hypoxic-ischemic brain damage in the rat. *Ann. Neurol.* **1981**, *9*, 131–141. [CrossRef]

16. Bona, E.; Hagberg, H.; Løberg, E.M.; Bågenholm, R.; Thoresen, M. Protective effects of moderate hypothermia after neonatal hypoxia-ischemia: Short- and long-term outcome. *Pediatr. Res.* **1998**, *43*, 738–745. [CrossRef]

17. Sabir, H.; Scull-Brown, E.; Liu, X.; Thoresen, M. Immediate hypothermia is not neuroprotective after severe hypoxia-ischemia and is deleterious when delayed by 12 hours in neonatal rats. *Stroke* **2012**, *43*, 3364–3370. [CrossRef]

18. Rodríguez-Fanjul, J.; Fernández-Feijóo, C.D.; Lopez-Abad, M.; Ramos, M.G.L.; Caballé, R.B.; Alcántara-Horillo, S.; Camprubí, M.C. Neuroprotection with hypothermia and allopurinol in an animal model of hypoxic-ischemic injury: Is it a gender question? *PLoS ONE* **2017**, *12*, e0184643. [CrossRef]

19. Rodríguez-Fanjul, J.; Fernández-Feijóo, C.D.; Camprubí, M.C. A new technique for collection of cerebrospinal fluid in rat pups. *J. Exp. Neurosci.* **2015**, *9*, 37–41. [CrossRef]

20. Benzie, I.F.F.; Strain, J.J. The ferric reducing ability of plasma (FRAP) as a measure of "antioxidant power": The FRAP assay. *Anal. Biochem.* **1996**, *239*, 70–76. [CrossRef]

21. Re, R.; Pellegrini, N.; Proteggente, A.; Pannala, A.; Yang, M.; Rice-Evans, C. Antioxidant activity applying an improved ABTS radical cation decolorization assay. *Free Radic. Biol. Med.* **1999**, *26*, 1231–1237. [CrossRef]

22. González-Diez, B.; Cavia, M.; Torres, G.; Abaigar, P.; Muñiz, P. Effect of a hemodiafiltration session with on-line regeneration of the ultrafiltrate on oxidative stress. *Blood Purif.* **2008**, *26*, 505–510. [CrossRef] [PubMed]

23. Del Pino-García, R.; Gerardi, G.; Rivero-Pérez, M.D.; González-San José, M.L.; García-Lomillo, J.; Muñiz, P. Wine pomace seasoning attenuates hyperglycaemia-induced endothelial dysfunction and oxidative damage in endothelial cells. *J. Funct. Foods* **2016**, *22*, 431–445. [CrossRef]

24. Levine, R.L.; Garland, D.; Oliver, C.N.; Amici, A.; Climent, I.; Lenz, A.-G.; Ahn, B.-W.; Shaltiel, S.; Stadtman, E.R. Determination of carbonyl content in oxidatively modified proteins. *Methods Enzymol.* **1990**, *186*, 464–478. [CrossRef]

25. Robertson, N.J.; Faulkner, S.; Fleiss, B.; Bainbridge, A.; Andorka, C.; Price, D.; Powell, E.; Lecky-Thompson, L.; Thei, L.; Chandrasekaran, M.; et al. Melatonin augments hypothermic neuroprotection in a perinatal asphyxia model. *Brain* **2013**, *136*, 90–105. [CrossRef] [PubMed]

26. Aly, H.; Elmahdy, H.; El-Dib, M.; Rowisha, M.; Awny, M.; Elgohary, T.; Elbatch, M.; Hamisa, M.; El-Mashad, A.-R. Melatonin use for neuroprotection in perinatal asphyxia: A randomized controlled pilot study. *J. Perinatol.* **2015**, *35*, 186–191. [CrossRef]

27. Jatana, M.; Singh, I.; Singh, A.K.; Jenkins, D. Combination of systemic hypothermia and N-acetylcysteine attenuates hypoxic-ischemic brain injury in neonatal rats. *Pediatr. Res.* **2006**, *59*, 684–689. [CrossRef]

28. Maiwald, C.A.; Annink, K.V.; Rüdiger, M.; Benders, M.J.N.L.; van Bel, F.; Allegaert, K.; Naulaers, G.; Bassler, D.; Klebermaß-Schrehof, K.; Vento, M.; et al. Effect of allopurinol in addition to hypothermia treatment in neonates for hypoxic-ischemic brain injury on neurocognitive outcome (ALBINO): Study protocol of a blinded randomized placebo-controlled parallel group multicenter trial for superiority (phase III). *BMC Pediatr.* **2019**, *27*, 210. [CrossRef]

29. Zhao, P.; Zhou, R.; Li, H.N.; Yao, W.X.; Qiao, H.Q.; Wang, S.J.; Niu, Y.; Sun, T.; Li, Y.X.; Yu, J.Q. Oxymatrine attenuated hypoxic-ischemic brain damage in neonatal rats via improving antioxidant enzyme activities and inhibiting cell death. *Neurochem. Int.* **2015**, *89*, 17–27. [CrossRef]

30. Wei, W.; Lan, X.-B.; Liu, N.; Yang, J.-M.; Du, J.; Ma, L.; Zhang, W.-J.; Niu, J.-G.; Sun, T.; Yu, J.-Q. Echinacoside alleviates hypoxic-ischemic brain injury in neonatal rat by enhancing antioxidant capacity and inhibiting apoptosis. *Neurochem. Res.* **2019**, *44*, 1582–1592. [CrossRef]

31. Dirnagl, U.; Lindauer, U.; Them, A.; Schreiber, S.; Pfister, H.-W.; Koedel, U.; Reszka, R.; Freyer, D.; Villringer, A. Global cerebral ischemia in the rat: Online monitoring of oxygen free radical production using chemiluminescence in vivo. *Br. J. Pharmacol.* **1995**, *15*, 929–940. [CrossRef]

32. Ono, T.; Tsuruta, R.; Fujita, M.; Aki, H.S.; Kutsuna, S.; Kawamura, Y.; Wakatsuki, J.; Aoki, T.; Kobayashi, C.; Kasaoka, S.; et al. Xanthine oxidase is one of the major sources of superoxide anion radicals in blood after reperfusion in rats with forebrain ischemia/reperfusion. *Brain Res.* **2009**, *1305*, 158–167. [CrossRef]

33. Palmer, C.; Towfighi, J.; Roberts, R.L.; Heitjan, D.F. Allopurinol administered after inducing hypoxia-ischemia reduces brain injury in 7-day-old rats. *Pediatr. Res.* **1993**, *33*, 405–411. [CrossRef]

34. Kumar, A.; Mittal, R.; Khanna, H.D.; Basu, S. Free radical injury and blood-brain barrier permeability in hypoxic-ischemic encephalopathy. *Pediatrics* **2008**, *122*, e722–e727. [CrossRef]

35. Nagel, S.; Su, Y.; Horstmann, S.; Heiland, S.; Gardner, H.; Koziol, J.; Martinez-Torres, F.J.; Wagner, S. Minocycline and hypothermia for reperfusion injury after focal cerebral ischemia in the rat—Effects on BBB breakdown and MMP expression in the acute and subacute phase. *Brain Res.* **2008**, *1188*, 198–206. [CrossRef]

36. Kaandorp, J.J.; Benders, M.J.; Schuit, E.; Rademaker, C.M.; Oudijk, M.A.; Porath, M.M.; Oetomo, S.B.; Wouters, M.G.; van Elburg, R.; Franssen, M.T.; et al. Maternal allopurinol administration during suspected fetal hypoxia: A novel neuroprotective intervention? A multicentre randomised placebo controlled trial. *Arch. Dis. Child.-Fetal Neonatal Ed.* **2015**, *100*, F216–F223. [CrossRef] [PubMed]

37. Martin, L.J.; Brambrink, A.M.; Price, A.C.; Kaiser, A.; Agnew, D.M.; Ichord, R.N.; Traystman, R.J. Neuronal death in newborn striatum after hypoxia-ischemia is necrosis and evolves with oxidative stress. *Neurobiol. Dis.* **2000**, *7*, 169–191. [CrossRef] [PubMed]
38. Jenkinson, S.G.; Roberts, R.J.; Delemos, R.A.; Lawrence, R.A.; Coalson, J.J.; King, R.J.; Null, J.D.M.; Gerstmann, D.R. Allopurinol-induced effects in premature baboons with respiratory distress syndrome. *J. Appl. Physiol.* **1991**, *70*, 1160–1167. [CrossRef] [PubMed]
39. Alataş, O.; Sahin, A.; Colak, O.; Inal, M.; Köken, T.; Yaşar, B.; Karahüseyinoglu, E. Beneficial effects of allopurinol on glutathione levels and glutathione peroxidase activity in rat ischaemic acute renal failure. *J. Int. Med. Res.* **1996**, *24*, 33–39. [CrossRef]
40. Okatani, Y.; Wakatsuki, A.; Kaneda, C. Melatonin increases activities of glutathione peroxidase and superoxide dismutase in fetal rat brain. *J. Pineal Res.* **2000**, *28*, 89–96. [CrossRef] [PubMed]
41. Zhang, H.; Zhang, J.J.; Mei, Y.W.; Sun, S.G.; Tong, E.T. Effects of immediate and delayed mild hypothermia on endogenous antioxidant enzymes and energy metabolites following global cerebral ischemia. *Chin. Med. J.* **2011**, *124*, 2764–2766. [PubMed]
42. Zhao, H.; Chen, Y. Effects of mild hypothermia therapy on the levels of glutathione in rabbit blood and cerebrospinal fluid after cardiopulmonary resuscitation. *Iran. J. Basic Med. Sci.* **2015**, *18*, 194–198.
43. Mueller-Burke, D.; Koehler, R.C.; Martin, L.J. Rapid NMDA receptor phosphorylation and oxidative stress precede striatal neurodegeneration after hypoxic ischemia in newborn piglets and are attenuated with hypothermia. *Int. J. Dev. Neurosci.* **2008**, *26*, 67–76. [CrossRef]
44. Ni, X.; Yang, Z.-J.; Carter, E.L.; Martin, L.J.; Koehler, R.C. Striatal neuroprotection from neonatal hypoxia-ischemia in piglets by antioxidant treatment with EUK-134 or edaravone. *Dev. Neurosci.* **2011**, *33*, 299–311. [CrossRef]
45. Negro, S.; Benders, M.J.; Tataranno, M.L.; Coviello, C.; de Vries, L.S.; van Bel, F.; Groenendaal, F.; Longini, M.; Proietti, F.; Belvisi, E.; et al. Early prediction of hypoxic-ischemic brain injury by a new panel of biomarkers in a population of term newborns. *Oxidative Med. Cell. Longev.* **2018**, *2018*, 1–10. [CrossRef] [PubMed]
46. Lafuente, H.; Pazos, M.R.; Alvarez, A.; Mohammed, N.; Santos, M.; Arizti, M.; Alvarez, F.J.; Martinez-Orgado, J. Effects of cannabidiol and hypothermia on short-term brain damage in new-born piglets after acute hypoxia-ischemia. *Front. Neurosci.* **2016**, *10*, 323. [CrossRef] [PubMed]
47. Duncan, J.G.; Ravi, R.; Stull, L.B.; Murphy, A.M. Chronic xanthine oxidase inhibition prevents myofibrillar protein oxidation and preserves cardiac function in a transgenic mouse model of cardiomyopathy. *Am. J. Physiol. Circ. Physiol.* **2005**, *289*, H1512–H1518. [CrossRef] [PubMed]
48. Signorini, C.; Ciccoli, L.; Leoncini, S.; Carloni, S.; Perrone, S.; Comporti, M.; Balduini, W.; Buonocore, G. Free iron, total F2-isoprostanes and total F4-neuroprostanes in a model of neonatal hypoxic-ischemic encephalopathy: Neuroprotective effect of melatonin. *J. Pineal Res.* **2009**, *46*, 148–154. [CrossRef]
49. Bredemeier, M.; Lopes, L.M.; Eisenreich, M.A.; Hickmann, S.; Bongiorno, G.K.; D'Avila, R.; Morsch, A.L.B.; Stein, F.D.S.; Campos, G.G.D. Xanthine oxidase inhibitors for prevention of cardiovascular events: A systematic review and meta-analysis of randomized controlled trials. *BMC Cardiovasc. Disord.* **2018**, *18*, 1–11. [CrossRef] [PubMed]
50. Huun, M.U.; Garberg, H.T.; Escobar, J.; Chafer, C.; Vento, M.; Holme, I.M.; Saugstad, O.D.; Solberg, R. DHA reduces oxidative stress following hypoxia-ischemia in newborn piglets: A study of lipid peroxidation products in urine and plasma. *J. Perinat. Med.* **2017**, *46*, 209–217. [CrossRef]
51. Huun, M.U.; Garberg, H.T.; Buonocore, G.; Longini, M.; Belvisi, E.; Bazzini, F.; Proietti, F.; Saugstad, O.D.; Solberg, R. Regional differences of hypothermia on oxidative stress following hypoxia-ischemia: A study of DHA and hypothermia on brain lipid peroxidation in newborn piglets. *J. Perinat. Med.* **2018**, *47*, 82–89. [CrossRef] [PubMed]

antioxidants

Article

NAC and Vitamin D Restore CNS Glutathione in Endotoxin-Sensitized Neonatal Hypoxic-Ischemic Rats

Lauren E. Adams [1,†], Hunter G. Moss [2,†], Danielle W. Lowe [3,†], Truman Brown [2], Donald B. Wiest [4], Bruce W. Hollis [1], Inderjit Singh [1] and Dorothea D. Jenkins [1,*]

[1] Department of Pediatrics, 10 McLellan Banks Dr, Medical University of South Carolina, Charleston, SC 29425, USA; adamsla@musc.edu (L.E.A.); hollisb@musc.edu (B.W.H.); singhi@musc.edu (I.S.)

[2] Center for Biomedical Imaging, Department of Radiology, Medical University of South Carolina, 68 President St. Room 205, Charleston, SC 29425, USA; mossh@musc.edu (H.G.M.); brotrr@musc.edu (T.B.)

[3] Department of Psychiatry, Medical University of South Carolina, 67 Presidents St., MSC 861, Charleston, SC 29425, USA; clarkdw@musc.edu

[4] Department of Pharmacy and Clinical Sciences, College of Pharmacy, Medical University of South Carolina, Charleston, SC 29425, USA; wiestdb@musc.edu

* Correspondence: jenkd@musc.edu; Tel.: +1-843-792-2112

† Three first authors contributed equally to this work.

Abstract: Therapeutic hypothermia does not improve outcomes in neonatal hypoxia ischemia (HI) complicated by perinatal infection, due to well-described, pre-existing oxidative stress and neuroinflammation that shorten the therapeutic window. For effective neuroprotection post-injury, we must first define and then target CNS metabolomic changes immediately after endotoxin-sensitized HI (LPS-HI). We hypothesized that LPS-HI would acutely deplete reduced glutathione (GSH), indicating overwhelming oxidative stress in spite of hypothermia treatment in neonatal rats. Post-natal day 7 rats were randomized to sham ligation, or severe LPS-HI (0.5 mg/kg 4 h before right carotid artery ligation, 90 min 8% O_2), followed by hypothermia alone or with N-acetylcysteine (25 mg/kg) and vitamin D (1,25(OH)$_2$D$_3$, 0.05 µg/kg) (NVD). We quantified in vivo CNS metabolites by serial 7T MR Spectroscopy before, immediately after LPS-HI, and after treatment, along with terminal plasma drug concentrations. GSH was significantly decreased in all LPS-HI rats compared with baseline and sham controls. Two hours of hypothermia alone did not improve GSH and allowed glutamate + glutamine (GLX) to increase. Within 1 h of administration, NVD increased GSH close to baseline and suppressed GLX. The combination of NVD with hypothermia rapidly improved cellular redox status after LPS-HI, potentially inhibiting important secondary injury cascades and allowing more time for hypothermic neuroprotection.

Keywords: glutathione; glutamate; oxidative stress; hypoxia ischemia; endotoxin; magnetic resonance spectroscopy; N-acetylcysteine; vitamin D

Citation: Adams, L.E.; Moss, H.G.; Lowe, D.W.; Brown, T.; Wiest, D.B.; Hollis, B.W.; Singh, I.; Jenkins, D.D. NAC and Vitamin D Restore CNS Glutathione in Endotoxin-Sensitized Neonatal Hypoxic-Ischemic Rats. *Antioxidants* **2021**, *10*, 489. https:// doi.org/10.3390/antiox10030489

Academic Editors: Julia Kuligowski and Máximo Vento

Received: 17 February 2021
Accepted: 17 March 2021
Published: 20 March 2021

Publisher's Note: MDPI stays neutral with regard to jurisdictional claims in published maps and institutional affiliations.

1. Introduction

Intrauterine inflammation and infection induce fetal inflammation, sensitize the fetal brain to secondary hypoxic ischemic (HI) injury, and increase the severity of HI brain injury [1–4]. Chorioamnionitis and/or funisitis are present in the placentas of approximately 30% of infants with hypoxic ischemic encephalopathy (HIE) [5]. Hypothermia treatment is the standard of care in moderate to severe HIE, including neonates exposed to intrauterine inflammation. While therapeutic hypothermia provides effective neuroprotection in both animals and neonates in uncomplicated HI, there is inconsistent benefit when infection/inflammation precedes HI [6–9]. As perinatal infection may not be identified prior to delivery [10], adjuvant postnatal therapies that improve on hypothermia's neuroprotection should also address the combined injury of inflammation and HI [5].

Few therapies tested in preclinical HI or lipopolysaccharide-sensitized HI (LPS-HI) models show synergetic effects with hypothermia [11–16]. While this may be due to the fact that therapeutic hypothermia already encompasses multiple mechanisms of neuroprotection, it may also be the case that pre-existing neuroinflammation requires much more rapid and targeted treatment of specific vulnerabilities [8], such as reduced glutathione (GSH) depletion. Fetal neuroinflammation in utero activates toll-like receptors, pro-inflammatory cytokines and oxidative stress, decreasing cellular reserves prior to HI [17,18] and leaving the neonate unable to counteract further hypoxic ischemic challenges at birth [19]. If oxidative stress and neuroinflammatory cascades are already activated before HI, the therapeutic window before secondary ATP depletion and irreversible injury is much shorter than in uncomplicated HI. In a normothermic neonatal rat, LPS-HI decreases GSH as soon as 2 h after HI [17]. What is not known is how rapidly CNS metabolomics are affected by LPS-HI and if hypothermia or other targeted treatments can act quickly enough within this 2 h time period to change the pro-oxidant CNS milieu in order to mitigate secondary injury.

Neutralizing oxidative stress early after injury is an important step in neuroprotective strategies. Rescuing GSH is necessary and sufficient for neuroprotection in glutamate toxicity, stroke and other injury models [20–27]. Antioxidant treatment with N-acetylcysteine (NAC) scavenges oxidative free radicals and provides cysteine, a critical precursor for GSH synthesis. In an LPS-HI neonatal model, NAC at 200 mg/kg increased GSH when given before and after HI under normothermic conditions [17]. While prior treatment with NAC is neuroprotective in normothermic neonatal LPS models, we have not determined whether NAC can act quickly when administered after LPS-HI to counteract the abnormalities of pre-existing neuroinflammation and rescue GSH.

We therefore designed these experiments to determine the hyperacute effects of LPS-HI injury on CNS metabolomics, and if treatment with therapeutic hypothermia alone or with a combination of antioxidant therapies could mitigate these during a critical early stage of secondary injury. We used serial magnetic resonance spectroscopy (MRS) measurements in the same animals before and after injury to analyze responses within individual animals, and to be able to draw conclusions in spite of the significant heterogeneity inherent in neonatal HI models, which is also highly relevant to the clinical disease process. We hypothesized that LPS-HI injury would rapidly decrease GSH in the ipsilateral hemisphere, and that hypothermia treatment alone would not mitigate this oxidative stress.

We tested the combination of NAC and $1,25(OH)_2D$ (NVD) with hypothermia, as we have previously shown that they are neuroprotective postnatally in both sexes in a neonatal model of severe HI [28]. NAC and $1,25(OH)_2D$ may act synergistically to increase the intracellular antioxidant capacity and decrease inflammation: NAC provides the rate-limiting glutathione substrate while vitamin D induces GSH synthetic enzymes cysteine-glutamate ligase and glutathione reductase [29,30]. By augmenting both aspects of GSH synthesis, we postulated that NVD treatment might be more effective at replenishing CNS GSH rapidly in the early phase of LPS-HI injury than hypothermia alone. We related these CNS metabolite changes to plasma concentrations of NAC and $1,25(OH)_2D$ for translational evidence that NVD crosses the blood–brain barrier quickly with significant therapeutic effects on CNS metabolomics, facilitating hypothermia's neuroprotective effects.

2. Materials and Methods

2.1. Validation of GSH by MRS

Prior to animal experimentation, we used VeSPA [31], a spectral simulation program, to create a custom simulated echo acquisition mode (STEAM) basis set for use at echo time (TE) 3 ms on 7 Tesla MRI that included GSH and all standard metabolites of total choline (tCho, choline + phosphocholine), total creatine (tCr, creatin + phosphocreatine), total N-Acetylaspartate (NAA, N-Acetylaspartate + N-acetylaspartylglutamate), glutamate (Glu), glutamine (Gln), glutamate + glutamine (GLX), taurine (Tau) and inositol (Ins). The basis set was calibrated and then imported into the LCModel for automatic spectral fitting of the data [32]. Phantom solutions (0.5–5 mM GSH, 5 mM dithiothreitol, 10 mM choline, 25 mM creatine, phosphate buffered saline, pH 7.1) were used to validate GSH, creatine and choline quantification by MRS, using the water peak as a standard. The quantification

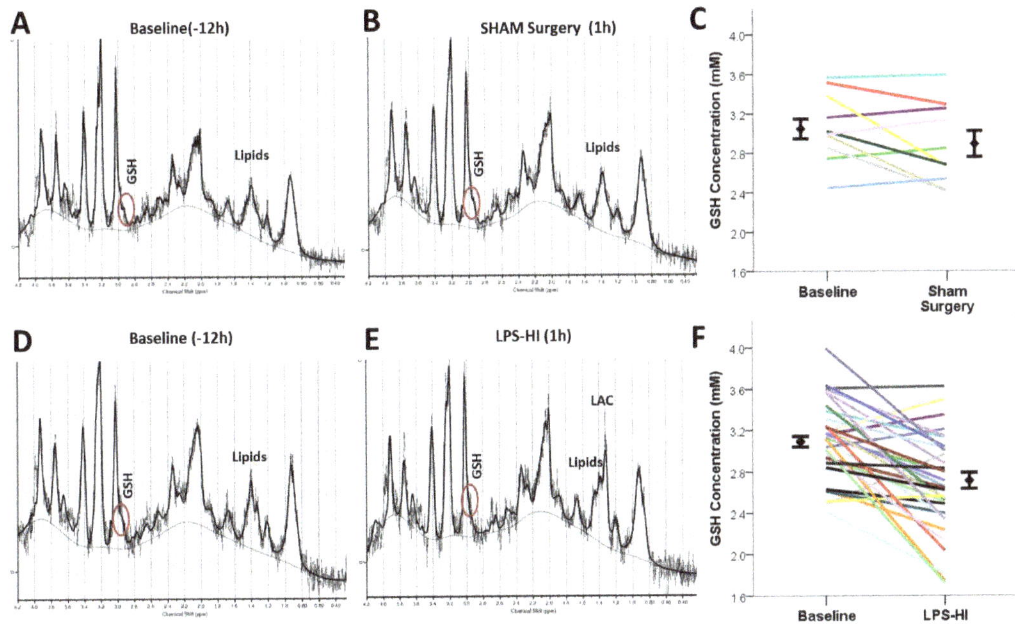

Figure 3. Representative spectra and change in GSH over time from sham (**A–C**) and LPS-HI rat pups (**D–F**). Lactate peaks are minimal in the sham animals before (**A**) and after sham surgery (**B**) and cannot be distinguished from lipid peaks in the MRS spectra. (**C**) Sham animals show no significant change in GSH before and after sham surgery ($n = 10$). (**D,E**) Lactate peak increases significantly after LPS-HI injury and is clearly visible in spectra from a representative rat (**E**). (**F**) LPS-HI animals show a significant decrease in GSH concentrations after LPS-HI injury ($p < 0.00017$, $n = 37$). Group means and standard errors are noted. Table 1. LPS-HI effects on MRS metabolite concentrations (mM, mean, SE) from PRE (PND 6) to POST-HI (PND 7 sham, or after LPS-HI). p values are noted for pairwise comparisons by scan time within LPS-HI group by generalized linear mixed model (GLMM) with Bonferroni adjustments, controlling for sex. The number of adequate spectra for each metabolite varies depending on exclusion due to Cramer Rao bounds >15%.

Table 1. LPS-HI effects on MRS metabolite concentrations (mM, estimated mean, SE) from PRE (PND 6) to POST-HI (PND 7 sham, or after LPS-HI). p values are noted for pairwise comparison (*) by scan time within LPS-HI group by GLMM with Bonferroni adjustments, controlling for sex. The number of adequate spectra for each metabolite varies depending on exclusion due to Cramer Rao bounds > 15%. (Sham group: PRE PND 6 $n = 12$, PND 7 $n = 10$; LPS-HI group: PRE PND 6 $n = 49$–52, POST HI $n = 36$–37).

	GSH		GLX		tCr		NAA		tCho		Ins		Tau	
	SHAM	LPS-HI	SHAM	LPS-HI	SHAM	LPS-HI	SHAM	LPS-HI	SHAM	LPS-HI	SHAM	LPS-HI	SHAM	LPS-HI
PRE PND 6	3.06 ± 0.11	3.06 ± 0.19 *	9.08 ± 0.19	9.00 ± 0.39 *	8.64 ± 0.17	8.58 ± 0.22	4.98 ± 0.19	4.75 ± 0.20	3.02 ± 0.07	2.89 ± 0.17	3.20 ± 0.42	3.13 ± 0.98	17.22 ± 0.48	16.70 ± 1.08
PND 7/ POST HI	2.89 ± 0.11	2.65 ± 0.19 *	9.57 ± 0.20	7.59 ± 0.40 *	8.44 ± 0.18	8.41 ± 023	4.80 ± 0.09	5.02 ± 0.10	3.01 ± 0.07	2.71 ± 0.18	2.95 ± 0.57	3.15 ± 0.90	16.25 ± 0.54	16.18 ± 1.6
p value	ns	<0.0001	ns	<0.0001	ns	ns	ns	ns	ns	<0.001	ns	ns	ns	ns

Sham animals showed no significant change in any metabolite after receiving the same anesthesia, surgery and scan times as LPS-HI rats. There was no difference at baseline scan (PRE) between sham and LPS-HI rats for any metabolite (Table 1).

3.3. NVD Improves GSH after LPS-HI Compared with HYPO Alone

In the VEH group that received HYPO alone, GSH decreased further from POST HI to POST HYPO and was significantly lower than PRE after hypothermia alone (overall model VEH n = 15, F = 5.2, p = 0.007, Figure 4A, Table 2). In contrast, GSH improved significantly after NVD at the POST HYPO scan compared with POST HI and was not significantly different from PRE GSH (overall model n = 20, F = 17.1, p = 0.001; Figure 4B, Table 2). The change in GSH from POST HI to POST HYPO was significantly less with hypothermia alone (ΔGSH VEH -0.08 ± 0.11 mM) than with hypothermia and NVD (ΔGSH $+0.32 \pm 0.1$ mM, p = 0.01, Figure 4C). GSH was not significantly different at any time point by sex.

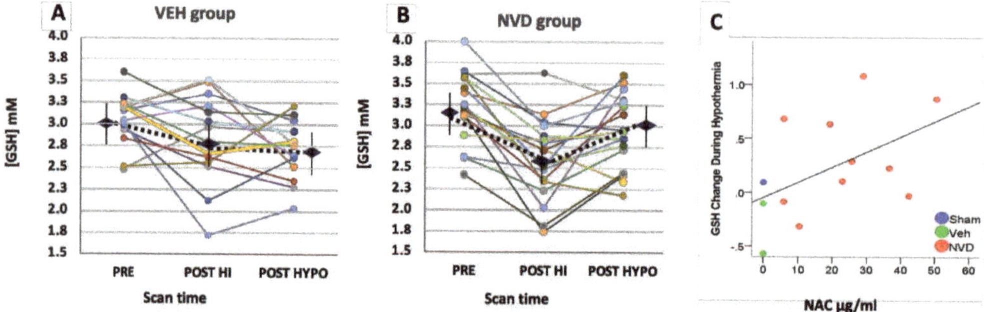

Figure 4. GSH concentrations in individual LPS-HI rats with adequate spectra for all 3 scans by VEH (**A**) or NVD (**B**) treatment over time and compared with NAC plasma concentrations (**C**). (**A**,**B**) Individual changes in GSH over serial scans at baseline (PRE), after LPS-HI (POST HI), and again immediately after 2 h of hypothermia (POST HYPO) in rats with VEH (**A**) or NVD (**B**) treatment. The markers show group mean and SE for each scan time. (**A**) The majority of VEH animals treated with saline do not show significant recovery with hypothermia alone and mean GSH continues to decrease at the third scan (n = 14). (**B**) In contrast, NVD treatment results in significant increases in mean GSH with hypothermia (n = 18). (**C**) Individual plasma NAC concentrations (n = 13) sampled immediately after last scan (POST HYPO) versus the change in GSH from immediately before and after NVD/saline. Sham concentration is included for reference.

Table 2. NVD vs. saline effects on MRS metabolite concentrations (mM, mean, standard error over 3 scans PRE, POST HI and POST HYPO, in the LPS-HI group treated with hypothermia. Significant changes between scan times within treatment groups are indicated by symbols, along with p values for the comparisons (* or ^) by GLMM with sequential Bonferroni adjustments, controlling for sex. The number of adequate spectra for each metabolite varies depending on exclusion due to Cramer Rao bounds >15%.

	GSH		GLX		tCr		NAA		tCho		Ins		Tau	
	VEH	NVD	VEH	NVD	VEH	NVD	VEH	NVD	VEH	NVD	VEH	NVD	VEH	NVD
PRE PND 6	2.97 ± 0.22 *^	3.16 ± 0.32 *^	8.94 ± 0.26 *^	9.02 ± 0.74 *^	8.65 ± 0.27	8.52 ± 0.35	4.96 ± 0.12	5.14 ± 0.13	2.82 ± 0.24	2.97 ± 0.25 *^	3.32 ± 0.91	3.16 ± 0.91	16.70 ± 1.08	16.18 ± 1.60
POST HI	2.72 ± 0.22 ^	2.58 ± 0.32 *	7.41 ± 0.28 *	7.77 ± 0.74 *	8.37± 0.28 *	8.44 ± 0.35	4.74 ± 0.19	4.94 ± 0.16	2.66 ± 0.24	2.75 ± 0.25 *	2.86 ± 0.92	2.85 ± 0.91	15.85 ± 1.10	15.74 ± 1.62
POST HYPO	2.65 ± 0.22 *	2.90 ± 0.32 *^	8.16 ± 0.29 *^	7.54 ± 0.74 ^	8.96 ± 0.28 *	8.67 ± 0.35	4.94 ± 0.16	5.14 ± 0.13	2.71 ± 0.24	2.72 ± 0.25 ^	3.42 ± 0.92	2.83 ± 0.91	16.45 ± 1.10	15.59 ± 1.62
p value	<0.05	<0.01	<0.005	<0.0001	≤ 0.01	ns	ns	ns	ns	<0.005	ns	ns	ns	ns

3.4. NAC and Active Vitamin D Increase GSH Response Rate in Heterogeneous LPS-HI Injury

VEH animals showed no improvement in mean GSH as a result of saline and hypothermia (p = 0.4, Figure 4A, Table 1). However, several individual VEH rats (3/14, 21%, Figure 4A) did respond to hypothermia alone with an increase in GSH. At the same time,

NVD increased GSH in 14/18 (78%) rat pups, which is significantly greater than in the VEH group (p = 0.004, fisher's exact test; Figure 4B). The plasma NAC concentrations obtained 1.5 h after i.p. injection of NVD/saline were positively correlated with the change in the CNS GSH from POST HI to POST HYPO (Spearman's rho = 0.552, p = 0.05, Figure 4C).

3.5. NAC and Active Vitamin D Suppress Glutamate + Glutamine after LPS-HI Injury

GLX decreased over the three scans in both LPS-HI groups (both F = 27, p < 0.0001). However, the VEH group exhibited a significant rebound in GLX after hypothermia alone, primarily due to increased glutamate (ΔGLX +0.76 $\pm$ 0.24 mM from PRE, F = 27, p = 0.002), whereas NVD continued to suppress glutamate (ΔGLX −0.23 $\pm$ 0.21 mM) from the second scan POST HI to the third scan after NVD and hypothermia (Table 2). In NVD rats, tCho decreased significantly with LPS-HI and remained lower at the third scan POST HYPO, while there are no substantive changes in tCho in the VEH group.

We found significant differences by sex in response to hypothermia treatment in GLX (F = 46.6, p = 0.01) and tCr (F = 4.5, p = 0.013). VEH females showed a rebound in CNS GLX and tCr concentrations, with significantly higher GLX (ΔGLX +0.89 $\pm$ 0.34 mM) and tCr (ΔtCr +0.76 $\pm$ 0.30 mM) after hypothermia treatment alone than males (Figure 5). NVD equalized this difference between males and females, and neither GLX nor tCr were significantly different between sexes at either POST HI or POST HYPO scans.

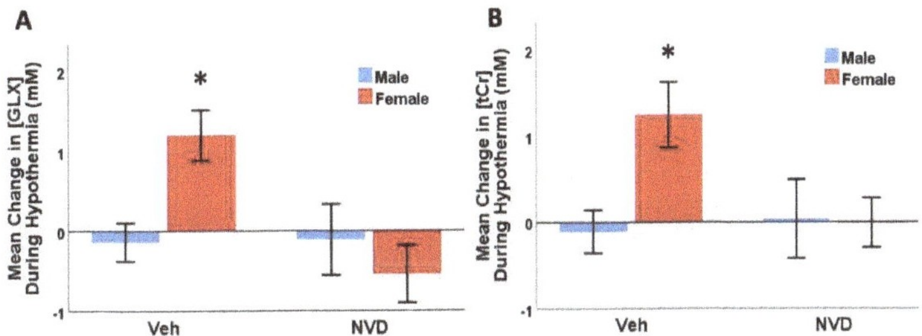

Figure 5. Changes in GLX (**A**) and tCr (**B**) between POST HI and POST HYPO scans by sex within treatment groups. VEH female animals (n = 6) showed a significant increase in mean GLX ((**A**), * p = 0.007), and Cr ((**B**), * p = 0.044) at 2.5 h after injury compared to VEH males (n = 9). There was no difference between females (n = 9) and males (n = 11) in the NVD group.

3.6. Lactate Increases after LPS-HI Injury and Persists after HYPO

Lactate was undetectable on PRE scans, but we observed significantly increased lactate peaks in the spectra from 22 of 37 animals (60%) obtained immediately after LPS-HI injury (Figure 2E). Lactate was quantifiable in POST HI scans with a mean SE of 4.34 $\pm$ 0.26 mM, and it persisted in 32% of pups in the POST HYPO period with mean SE 3.12 $\pm$ 0.26 mM. There was no difference in the numbers of animals that had detectable lactate in either NVD or VEH groups by chi-squared analysis. Spectral fitting yielded Cramer Rao standard deviations of 7–38%, consistent with greater difficulties fitting the lactate peak within overlapping lipid residues.

3.7. LPS-HI Injury Decreases Plasma 1,25(OH)$_2$D

LPS-HI was associated with a 50% decrease in mean circulating 1,25(OH)$_2$D concentrations (42 $\pm$ 21 pg/mL) compared with sham animals (84 $\pm$ 34 pg/mL, Figure 6A) in blood collected 2.5 h after hypoxia. Neither hypothermia alone nor NVD in combination with hypothermia restored plasma concentrations of active hormone 1,25(OH)$_2$D within this time period (1.5 h after NVD treatment). At the same time, all LPS-HI animals had slightly higher circulating 25(OH)D$_3$ (18.5 $\pm$ 2.6 ng/mL) than sham rats (14.9 $\pm$ 1.7 ng/mL, Figure 6B), indicating the increased utilization or degradation of vitamin D, and attempts to mobilize the inactive precursor to maintain active hormone levels. There were no differences by sex within or between the sham and LPS-HI groups.

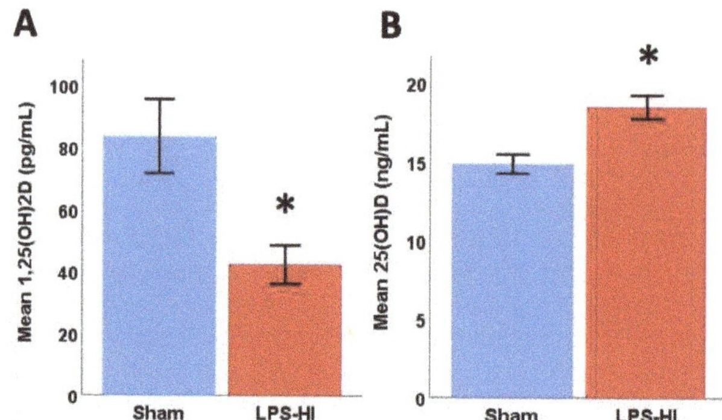

Figure 6. (**A**) Endotoxin-sensitized HI (LPS-HI) animals ($n = 11$), regardless of treatment, show a 50% lower plasma concentration of 1,25(OH)$_2$D after LPS-HI than sham animals ($n = 8$, * $p = 0.004$, *t*-test), (**B**) and marginally higher plasma 25(OH)D (+4 ng/mL, $n = 12$) than sham ($n = 8$, * $p = 0.002$, *t*-test).

3.8. T2 Evidence of Infarcts

T2 images demonstrated hyperacute infarction in four animals after hypothermia treatment, two in each of the VEH and NVD groups (Figure 7) within 2 h of hypoxia, which usually evolves over 24 h [37].

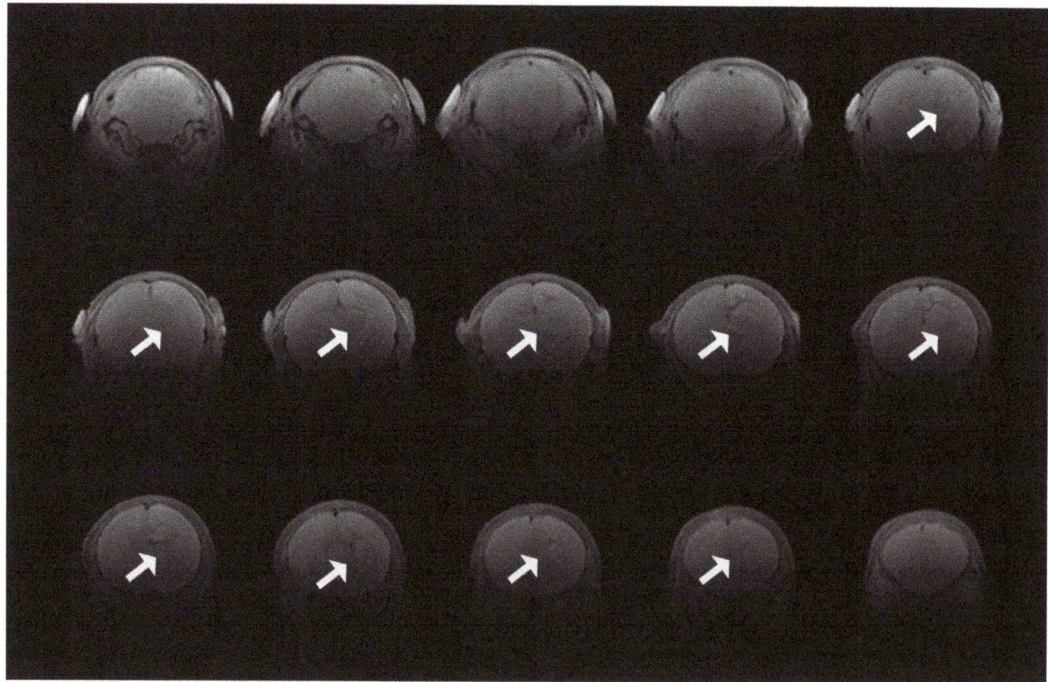

Figure 7. T2 scan showing progression of infarction for an animal from the VEH group. Of note, the right hemisphere is on the left side of the magnetic resonance imaging (MRI) images.

4. Discussion

Using serial in vivo 7T MRS, we show that endotoxin-sensitized HI injury significantly and immediately changes CNS redox status in a neonatal model, decreasing the intracellular antioxidant concentrations of reduced glutathione GSH in the affected hemisphere immediately after hypoxia, compared with pre-injury and with sham controls. The clinical standard of care for HIE, hypothermia, did not rescue mean GSH within 2 h and concentrations continued to decrease in the POST HYPO period. However, one dose of *N*-acetylcysteine (25 mg/kg) and calcitriol (0.05 mcg/kg), administered after 1 h of hypothermia, restored GSH in the affected hemisphere to near pre-injury levels POST HYPO in the NVD group. Further, 78% of NVD-treated rat pups showed a significant increase in GSH, while only 21% of rat pups treated with hypothermia alone had increased GSH. Plasma NAC positively correlated with CNS GSH, indicating that NAC rapidly crossed the blood–brain barrier to increase GSH [38].

Previous work in an LPS-HI neonatal rat model using biochemical assays of GSH agree with our in vivo MRS findings. In one investigation in neonatal rats, LPS prior to HI increased the depletion of GSH and increased F2-isoprostanes and peroxynitrite derivatives (nitrosylated proteins) at 2 h after HI compared with controls and LPS alone [17]. NAC treatment at 200 mg/kg prior to or immediately after LPS-HI provided neuroprotection, but not 25 mg/kg or when NAC administration was delayed after injury under normothermic conditions [17]. We tested NVD under hypothermic conditions, as hypothermia is now the standard of care for hypoxic ischemic injury in human neonates. In our severe HI and now LPS-HI neonatal rat models, we have shown that low-dose NAC (25 mg/kg) combined with active vitamin D and hypothermia improves neuroinflammation and provides neuroprotection (HI) [28], as well as facilitating facilitates increased glutathione and decreased glutamate (LPS-HI) over hypothermia alone. Taken together, the data suggest that hypothermia decreases the dose of NAC required to affect redox state and outcomes [17,28,39].

Reduced glutathione plays a central role in cell survival in acute HI and neuroinflammation and is an important outcome measure [40–44]. GSH is oxidized to the dimer GSSG by glutathione peroxidase after detoxifying superoxide and peroxynitrite free radicals [45]. The depletion of GSH triggers or potentiates glutamate- and nitric oxide-induced apoptosis [26,46,47]. If oxidative stress and GSH depletion continue, secondary ATP depletion and glutamate-induced cell death follows the initial repletion of ATP [20]. Restoring GSH, particularly in the mitochondria of astrocytes, is necessary and sufficient for neuroprotection from neuronal glutamate toxicity [20], and is essential for cell survival after multiple different insults [21–24,27,45,48,49].

NVD also mitigates the increase in GLX seen in the hypothermia alone LPS-HI rat pups. While resolving glutamate and glutamine peaks with confidence is controversial in MR spectroscopy, glutamate accounts for the majority of the GLX peak complex and GLX concentration. The depletion of GSH been shown to augment glutamate toxicity after HI, and increased extracellular glutamate inhibits cystine uptake and leads to further GSH depletion [48]. We found an increase in GLX and GSH depletion after hypothermia alone, supporting the relationship of lower GSH and increased glutamate toxicity with neuroinflammation in this severe LPS-HI model, as found by other investigators after HI alone [48]. Conversely, NVD resulted in an increase in GSH and decrease in glutamate concentrations, measured by GLX. Surprisingly, the increase in glutamate was seen primarily in the female rats in the vehicle group, rather than in males. Few reports have addressed glutamate toxicity as a sexually dimorphic injury pathway, but one such study showed that female PND7 rats had greater hippocampal injury than males after administration of the glutamate receptor agonist kainate [50]. However, males and females may have different time courses of glutamate, and our time point may have been too early to observe an increase in males.

The neurohormone 1,25(OH)$_2$D plays an important role in neuroprotection in animal models [28,51–56], as well as in human stroke and other CNS diseases [35,57–64]. Studies also indicate that inflammation results in the degradation of active vitamin D that may persist for weeks after injury [28,65]. At 11 days after severe HI in neonatal rats, inducible nitric oxide synthase expression in the CNS correlated strongly with the expression of the CYP27B1 enzymes responsible for vitamin D catabolism [28]. Here we report that active vitamin D is dramatically decreased in neonatal rat plasma 2.5 h after LPS-HI, much more quickly than previously demonstrated. These data add to our and others' work showing vitamin D deficiency and increased vitamin D degradation in neuroinflammatory conditions: in human HIE neonates [35,66], in animal models of HI [28], after antenatal endotoxin exposure [65], and in other conditions of immune activation [36]. These data indicate that vitamin D metabolism is disrupted in both HI and intrauterine inflammation.

Perhaps because of increased vitamin D catabolism with inflammation, we did not demonstrate an increase in plasma $1,25(OH)_2D$ concentration after the administration of calcitriol in LPS-HI, unlike a previous report in normal adult rats, in which serum $1,25(OH)_2D$ was increased one hour after i.p. administration [67]. This may be due to its rapid dispersal into tissues and increased utilization, or the metabolism of active hormones in inflammatory conditions [28,35]. In HIE infants, vitamin D-binding proteins and albumin that normally function to maintain circulating vitamin D concentrations and limit distribution into tissues are significantly decreased, even with hypothermia [35].

Active vitamin D is a neurohormone that regulates the transcription of multiple genes, including the GSH synthetic enzymes that are inhibited by acidosis [29,52,56,68]. In our work, NAC and active vitamin D acutely and significantly increase GSH in LPS-HI rats and in the basal ganglia on PND 5 after rewarming from hypothermia treatment, within 12–30 min after completing the infusion in HIE neonates [38]. Using biochemical assays, other investigations determined that NAC and active vitamin D increase GSH [29]. Similar to our strategy of using NVD combination therapy to increase intracellular GSH, other investigators have paired NAC with probencid in the treatment of pediatric traumatic brain injury to increase GSH synthesis and inhibit the transport of GSH out of the cell [24,69]. We did not investigate the effect of NVD synergy on GSH synthesis with our targeted study design in this proof-of-concept study. Improved intracellular redox status has been shown to augment $1,25(OH)_2D$ binding to its receptor in animal models, enhancing multiple vitamin D signaling transduction pathways [70]. It is also possible that the co-administration of NAC and vitamin D have synergistic effects on important anti-inflammatory and anti-apoptotic pathways other than GSH synthesis.

The use of MRS for measuring hyperacute changes in CNS metabolic profiles represents a non-invasive method for assessing the precise evolution of cellular injury [71,72]. The advantages of MRS quantification versus biochemical assays of reduced glutathione include the fact that oxidative stress may be measured serially, before and after injury and treatment in the same animals; antioxidant dosing can show definitive CNS target attainment; and the real-time in vivo measurement is not subject to rapid oxidation during tissue processing, as it is with biochemical assay. MRS is a non-invasive method of measuring GSH concentrations that has been validated using short echo STEAM sequences in human adults and neonates [73], and allows preclinical pharmacokinetic and pharmacodynamic studies to quickly translate effective doses of therapeutics to clinical studies. Newer software programs use MRS pattern recognition to classify infarct evolution in adult stroke models, illustrating the ability of serial metabolomics to discriminate between core and penumbra after HI [73]. In addition, MRS can quantify cellular energetics and redox simultaneously after CNS injury via measurement of GSH and tCr at the same echo time. This is an important feature of MRS, as the restoring of both are key therapeutic targets. While MR diffusion sequences have proven useful for identifying cytotoxic edema after HI injury [37,74,75], MRS offers significantly more nuanced and complex data on cellular state after injury and recovery, enabling CNS pharmacodynamic studies [76]. MRS metabolomics are valuable biomarkers that provide translationally important diagnostic and therapeutic information after stroke, particularly in the hyperacute phase when therapeutic intervention may improve outcomes.

The limitations of our study include that we did not test whether calcitriol alone could augment GSH production over NAC alone in this short time frame, and we did not perform MRS in the contralateral, LPS-hypoxia-only hemisphere, due to time and funding constraints with repeated MR scans on each rat, and the necessity of processing one litter in one day. Finally, GSSG is not measurable by MRS. However, GSH is the active antioxidant. Further, either GSH or GSSG may be consumed by glutathionylation, an important post-translational regulatory mechanism [48]. Therefore, GSSG concentration or GSH/GSSG ratios are perhaps less important given the current understanding that the synthesis of GSH–GSSG is not a closed loop, and the absolute GSH/GSSG ratio has little additional value for determining cellular ability to handle oxidative stress over GSH alone.

5. Conclusions

In one of the most severe models of CNS insult in neonates, serial acute MRS demonstrated significantly decreased concentrations of reduced CNS glutathione after LPS-HI, which were not rescued by hypothermia alone. However, low-dose NAC and active vitamin D administered one hour after the initiation of hypothermia resulted in rapid increases in GSH, which correlated with NAC plasma levels. Serial MRS in the hyperacute phase plays a unique role in translational studies by quantifying redox alterations after CNS injury via GSH and measuring a pharmacodynamic CNS marker of NAC and vitamin D effect that is associated with outcome and response to therapy after this severe injury.

Author Contributions: conceptualization, L.E.A., H.G.M., D.W.L., T.B., D.B.W., I.S. and D.D.J.; data curation, L.E.A., H.G.M., D.W.L., T.B., D.B.W., B.W.H. and D.D.J.; formal analysis, L.E.A., H.G.M., D.W.L., T.B., D.B.W., B.W.H. and D.D.J.; funding acquisition, H.G.M., T.B., I.S. and D.D.J.; investigation, L.E.A., H.G.M., D.W.L., T.B., D.B.W., B.W.H. and D.D.J.; methodology, H.G.M., D.W.L., T.B., D.B.W., B.W.H. and D.D.J.; project administration, D.W.L. and D.D.J.; resources, D.W.L., T.B., D.B.W., B.W.H., I.S. and D.D.J.; software, H.G.M., D.W.L. and D.B.W.; supervision, T.B., I.S. and D.D.J.; validation, H.G.M., D.W.L., T.B., D.B.W., B.W.H. and D.D.J.; visualization, T.B., D.B.W. and D.D.J.; writing—origiG.nal draft, L.E.A., D.W.L. and D.D.J.; writing—review and editing, L.E.A., H.G.M., D.W.L., T.B., D.B.W., B.W.H., I.S. and D.D.J. All authors have read and agreed to the published version of the manuscript.

Funding: This research was funded by the following grants: NIH NINDS F31NS108623 (Moss, H); BX003401 Department of Veteran Affairs (Singh, I).

Institutional Review Board Statement: All procedures were performed in accordance with the approved protocol #1678 by MUSC's Institutional Animal Care and Use Committee (approved and renewed on 3 year cycle, last 2/1/17) and the Guide for the Care and Use of Laboratory Animals adopted by the National Institutes of Health, and were approved by the MUSC Animal Care and Use Committee. We used the least number of animals required to ensure scientific rigor and minimized harm to the animals with anesthesia for experimental procedures.

Informed Consent Statement: Not Applicable.

Data Availability Statement: Data is contained within the article.

Acknowledgments: We wish to acknowledge the contributions of Andrew Barbour, who assisted with blood sampling and processing and euthanasia, and Xingju Nie who performed the MRS scans.

Conflicts of Interest: No author received payment to produce the manuscript nor has any conflict of interest to disclose.

Abbreviations

Cho	total choline (choline + phosphocholine)
tCr	total creatine (creatine + phosphocreatine)
GLMM	generalized linear mixed model
Gln	glutamine
Glu	glutamate

GLX	glutamate + glutamine
GSH	glutathione
GSSG	glutathione disulfide
HI	hypoxic ischemic
HIE	hypoxic ischemic encephalopathy
Ins	inositol
LAC	lactate
LPS-HI	lipopolysaccharide endotoxin-sensitized HI
MRS	magnetic resonance spectroscopy
NAA	total N-Acetylaspartate (N-Acetylaspartate + N-acetylaspartylglutamate)
NAC	N-acetylcysteine
NPM	N-(1-pyrenyl)maleimide
NVD	NAC and 1,25(OH)$_2$D
PND	postnatal day
POST HI	immediately after hypoxia
POST HYPO	immediately after hypothermia
STEAM	simulated echo acquisition mode
Tau	taurine
TE	echo time
VEH	saline

References

1. Barks, J.D.E.; Liu, Y.-Q.; Shangguan, Y.; Li, J.; Pfau, J.; Silverstein, F.S. Impact of indolent inflammation on neonatal hypoxic-ischemic brain injury in mice. *Int. J. Dev. Neurosci.* **2007**, *26*, 57–65. [CrossRef]
2. Eklind, S.; Mallard, C.; Leverin, A.-L.; Gilland, E.; Blomgren, K.; Mattsby-Baltzer, I.; Hagberg, H. Bacterial endotoxin sensitizes the immature brain to hypoxic-ischaemic injury. *Eur. J. Neurosci.* **2001**, *13*, 1101–1106. [CrossRef]
3. Garnier, Y.; Coumans, A.A.C.A.; Jensen, A.; Hasaart, T.H.M.; Berger, R. Infection-related perinatal brain injury: The pathogenic role of impaired fetal cardiovascular control. *J. Soc. Gynecol. Investig.* **2003**, *10*, 450–459. [CrossRef]
4. Martinello, K.A.; Meehan, C.; Avdic-Belltheus, A.; Lingam, I.; Ragab, S.; Hristova, M.; Tann, C.J.; Peebles, D.; Hagberg, H.; Wolfs, T.G.A.M.; et al. Acute LPS sensitization and continuous infusion exacerbates hypoxic brain injury in a piglet model of neonatal encephalopathy. *Sci. Rep.* **2019**, *9*, 1–17. [CrossRef] [PubMed]
5. Tann, C.J.; Martinello, K.A.; Sadoo, S.; Lawn, J.E.; Seale, A.C.; Vega-Poblete, M.; Russell, N.J.; Baker, C.J.; Bartlett, L.; Cutland, C.; et al. Neonatal Encephalopathy With Group B Streptococcal Disease Worldwide: Systematic Review, Investigator Group Datasets, and Meta-analysis. *Clin. Infect. Dis.* **2017**, *65*, S173–S189. [CrossRef] [PubMed]
6. Osredkar, D.; Thoresen, M.; Maes, E.; Flatebø, T.; Elstad, M.; Sabir, H. Hypothermia is not neuroprotective after infection-sensitized neonatal hypoxic–ischemic brain injury. *Resuscitation.* **2014**, *85*, 567–572. [CrossRef] [PubMed]
7. Jacobs, S.E.; Berg, M.; Hunt, R.; Tarnow-Mordi, W.O.; Inder, T.E.; Davis, P.G. Cooling for newborns with hypoxic ischaemic encephalopathy. *Cochrane Database Syst. Rev.* **2013**, *2013*, CD003311. [CrossRef]
8. Chevin, M.; Guiraut, C.; Maurice-Gelinas, C.; DesLauriers, J.; Grignon, S.; Sébire, G. Neuroprotective effects of hypothermia in inflammatory-sensitized hypoxic-ischemic encephalopathy. *Int. J. Dev. Neurosci.* **2016**, *55*, 1–8. [CrossRef]
9. Falck, M.; Osredkar, D.; Maes, E.; Flatebø, T.; Wood, T.R.; Sabir, H.; Thoresen, M. Hypothermic Neuronal Rescue from Infection-Sensitised Hypoxic-Ischaemic Brain Injury Is Pathogen Dependent. *Dev. Neurosci.* **2017**, *39*, 238–247. [CrossRef] [PubMed]
10. Wintermark, P.; Boyd, T.; Gregas, M.C.; Labrecque, M.; Hansen, A. Placental pathology in asphyxiated newborns meeting the criteria for therapeutic hypothermia. *Am. J. Obstet. Gynecol.* **2010**, *203*, 579.e1–579.e9. [CrossRef]
11. Martinello, K.; Meehan, C.; Avdic-Belltheus, A.; Mutshiya, T.; Yang, Q.; Price, D.; Sokolska, M.; Bainbridge, A.; Hristova, M.; Tann, C.J.; et al. Hypothermia is not neuroprotective in a piglet model of LPS sensitized neonatal encephalopathy. *J. Paediatr. Child Health* **2019**, *55*, 33. [CrossRef]
12. Lingam, I.; Meehan, C.; Avdic-Belltheus, A.; Martinello, K.; Hristova, M.; Kaynezhad, P.; Bauer, C.; Tachtsidis, I.; Golay, X.; Robertson, N.J. Short-term effects of early initiation of magnesium infusion combined with cooling after hypoxia–ischemia in term piglets. *Pediatr. Res.* **2019**, *86*, 699–708. [CrossRef]
13. Robertson, N.J.; Martinello, K.; Lingam, I.; Avdic-Belltheus, A.; Meehan, C.; Alonso-Alconada, D.; Ragab, S.; Bainbridge, A.; Sokolska, M.; Tachrount, M.; et al. Melatonin as an adjunct to therapeutic hypothermia in a piglet model of neonatal encephalopathy: A translational study. *Neurobiol. Dis.* **2019**, *121*, 240–251. [CrossRef]
14. Azzopardi, D.; Robertson, N.J.; Bainbridge, A.; Cady, E.; Charles-Edwards, G.; Deierl, A.; Fagiolo, G.; Franks, N.P.; Griffiths, J.; Hajnal, J.; et al. Moderate hypothermia within 6 h of birth plus inhaled xenon versus moderate hypothermia alone after birth asphyxia (TOBY-Xe): A proof-of-concept, open-label, randomised controlled trial. *Lancet Neurol.* **2016**, *15*, 145–153. [CrossRef]
15. Fang, A.Y.; Gonzalez, F.F.; Sheldon, R.A.; Ferriero, D.M. Effects of combination therapy using hypothermia and erythropoietin in a rat model of neonatal hypoxia–ischemia. *Pediatr. Res.* **2012**, *73*, 12–17. [CrossRef] [PubMed]

16. Chevin, M.; Guiraut, C.; Sébire, G. Effect of hypothermia on interleukin-1 receptor antagonist pharmacodynamics in inflammatory-sensitized hypoxic-ischemic encephalopathy of term newborns. *J. Neuroinflamm.* **2018**, *15*, 214. [CrossRef] [PubMed]

17. Wang, X.; Svedin, P.; Nie, C.; Lapatto, R.; Zhu, C.; Gustavsson, M.; Sandberg, M.; Karlsson, J.-O.; Romero, R.; Hagberg, H.; et al. N-acetylcysteine reduces lipopolysaccharide-sensitized hypoxic-ischemic brain injury. *Ann. Neurol.* **2007**, *61*, 263–271. [CrossRef] [PubMed]

18. Osredkar, D.; Sabir, H.; Falck, M.; Wood, T.K.; Maes, E.; Flatebø, T.; Puchades, M.; Thoresen, M. Hypothermia Does Not Reverse Cellular Responses Caused by Lipopolysaccharide in Neonatal Hypoxic-Ischaemic Brain Injury. *Dev. Neurosci.* **2015**, *37*, 390–397. [CrossRef]

19. Packer, L.; Weber, S.U.; Rimbach, G. Molecular Aspects of α-Tocotrienol Antioxidant Action and Cell Signalling. *J. Nutr.* **2001**, *131*, 369S–373S. [CrossRef]

20. Shih, A.Y.; Johnson, D.A.; Wong, G.; Kraft, A.D.; Jiang, L.; Erb, H.; Johnson, J.A.; Murphy, T.H. Coordinate Regulation of Glutathione Biosynthesis and Release by Nrf2-Expressing Glia Potently Protects Neurons from Oxidative Stress. *J. Neurosci.* **2003**, *23*, 3394–3406. [CrossRef]

21. Quintana-Cabrera, R.; Bolaños, J.P. Glutathione and γ-glutamylcysteine in the antioxidant and survival functions of mitochondria. *Biochem. Soc. Trans.* **2013**, *41*, 106–110. [CrossRef]

22. Marí, M.; Morales, A.; Colell, A.; García-Ruiz, C.; Fernández-Checa, J.C. Mitochondrial Glutathione, a Key Survival Antioxidant. *Antioxid. Redox Signal.* **2009**, *11*, 2685–2700. [CrossRef]

23. Armstrong, J.S.; Steinauer, K.K.; Hornung, B.; Irish, J.M.; Lecane, P.; Birrell, G.W.; Peehl, D.M.; Knox, S.J. Role of glutathione depletion and reactive oxygen species generation in apoptotic signaling in a human B lymphoma cell line. *Cell Death Differ* **2002**, *9*, 252–263. [CrossRef] [PubMed]

24. Du, L.; Empey, P.E.; Ji, J.; Chao, H.; Kochanek, P.M.; Bayır, H.; Clark, R.S. Probenecid and N-Acetylcysteine Prevent Loss of Intracellular Glutathione and Inhibit Neuronal Death after Mechanical Stretch Injury In Vitro. *J. Neurotrauma* **2016**, *33*, 1913–1917. [CrossRef] [PubMed]

25. Khan, M.; Sekhon, B.; Jatana, M.; Giri, S.; Gilg, A.G.; Sekhon, C.; Singh, I.; Singh, A.K. Administration of N-acetylcysteine after focal cerebral ischemia protects brain and reduces inflammation in a rat model of experimental stroke. *J. Neurosci. Res.* **2004**, *76*, 519–527. [CrossRef] [PubMed]

26. Li, Y.; Maher, P.; Schubert, D. A Role for 12-lipoxygenase in Nerve Cell Death Caused by Glutathione Depletion. *Neuron* **1997**, *19*, 453–463. [CrossRef]

27. Kirkland, R.A.; Franklin, J.L. Evidence for Redox Regulation of Cytochrome c Release during Programmed Neuronal Death: Antioxidant Effects of Protein Synthesis and Caspase Inhibition. *J. Neurosci.* **2001**, *21*, 1949–1963. [CrossRef]

28. Lowe, D.W.; Fraser, J.L.; Rollins, L.G.; Bentzley, J.; Nie, X.; Martin, R.; Singh, I.; Jenkins, D. Vitamin D improves functional outcomes in neonatal hypoxic ischemic male rats treated with N-acetylcysteine and hypothermia. *Neuropharmacology* **2017**, *123*, 186–200. [CrossRef]

29. Jain, S.K.; Micinski, D. Vitamin D upregulates glutamate cysteine ligase and glutathione reductase, and GSH formation, and decreases ROS and MCP-1 and IL-8 secretion in high-glucose exposed U937 monocytes. *Biochem. Biophys. Res. Commun.* **2013**, *437*, 7–11. [CrossRef]

30. Garcion, E.; Sindji, L.; Leblondel, G.; Brachet, P.; Darcy, F. 1,25-dihydroxyvitamin D3 regulates the synthesis of gamma-glutamyl transpeptidase and glutathione levels in rat primary astrocytes. *J. Neurochem.* **1999**, *73*, 859–866. [CrossRef]

31. Soher, B.J.; Semanchuk, P.; Todd, D.; Steinberg, J.; Young, K. Vespa: Integrated applications for RF pulse design, spectral simulation and MRS data analysis. *Proc. Int. Soc. Magn. Reson. Med.* **2011**, *19*, 1410.

32. Provencher, S.W. Estimation of metabolite concentrations from localized in vivo proton NMR spectra. *Magn. Reson. Med.* **1993**, *30*, 672–679. [CrossRef]

33. Kreis, R.; Hofmann, L.; Kuhlmann, B.; Boesch, C.; Bossi, E.; Hüppi, P. Brain metabolite composition during early human brain development as measured by quantitative in vivo1H magnetic resonance spectroscopy. *Magn. Reson. Med.* **2002**, *48*, 949–958. [CrossRef]

34. Wiest, D.B.; Chang, E.; Fanning, D.; Garner, S.; Cox, T.; Jenkins, D.D. Antenatal Pharmacokinetics and Placental Transfer of N-Acetylcysteine in Chorioamnionitis for Fetal Neuroprotection. *J. Pediatr.* **2014**, *165*, 672–677.e2. [CrossRef]

35. Lowe, D.W.; Hollis, B.W.; Wagner, C.L.; Bass, T.; A Kaufman, D.; Horgan, M.J.; Givelichian, L.M.; Sankaran, K.; Yager, J.Y.; Katikaneni, L.D.; et al. Vitamin D insufficiency in neonatal hypoxic–ischemic encephalopathy. *Pediatr. Res.* **2017**, *82*, 55–62. [CrossRef] [PubMed]

36. Paintlia, A.S.; Paintlia, M.K.; Hollis, B.W.; Singh, A.K.; Singh, I. Interference with RhoA-ROCK signaling mechanism in autoreactive CD4+ T cells enhances the bioavailability of 1,25-dihydroxyvitamin D3 in experimental autoimmune encephalomyelitis. *Am. J. Pathol.* **2012**, *181*, 993–1006. [CrossRef]

37. Bouts, M.J.R.J.; Westmoreland, S.V.; De Crespigny, A.J.; Liu, Y.; Vangel, M.; Dijkhuizen, R.M.; Wu, O.; D'Arceuil, H.E. Magnetic resonance imaging-based cerebral tissue classification reveals distinct spatiotemporal patterns of changes after stroke in non-human primates. *BMC Neurosci.* **2015**, *16*, 1–13. [CrossRef] [PubMed]

38. Moss, H.G.; Brown, T.R.; Wiest, D.B.; Jenkins, D.D. N-Acetylcysteine rapidly replenishes central nervous system glutathione measured via magnetic resonance spectroscopy in human neonates with hypoxic-ischemic encephalopathy. *Br. J. Pharmacol.* **2018**, *38*, 950–958. [CrossRef] [PubMed]

39. Nie, X.; Lowe, D.W.; Rollins, L.G.; Bentzley, J.; Fraser, J.L.; Martin, R.; Singh, I.; Jenkins, D. Sex-specific effects of N-acetylcysteine in neonatal rats treated with hypothermia after severe hypoxia-ischemia. *Neurosci. Res.* **2016**, *108*, 24–33. [CrossRef]

40. Du, L.; Hickey, R.W.; Bayir, H.; Watkins, S.C.; Tyurin, V.A.; Guo, F.; Kochanek, P.M.; Jenkins, L.W.; Ren, J.; Gibson, G.; et al. Starving Neurons Show Sex Difference in Autophagy. *J. Biol. Chem.* **2009**, *284*, 2383–2396. [CrossRef]

41. Mandal, P.K.; Saharan, S.; Tripathi, M.; Murari, G. Brain Glutathione Levels – A Novel Biomarker for Mild Cognitive Impairment and Alzheimer's Disease. *Biol. Psychiatry* **2015**, *78*, 702–710. [CrossRef] [PubMed]

42. An, L.; Zhang, Y.; Thomasson, D.M.; Latour, L.L.; Baker, E.H.; Shen, J.; Warach, S. Measurement of glutathione in normal volunteers and stroke patients at 3T using J-difference spectroscopy with minimized subtraction errors. *J. Magn. Reson. Imaging* **2009**, *30*, 263–270. [CrossRef] [PubMed]

43. Srinivasan, R.; Ratiney, H.; Hammond-Rosenbluth, K.E.; Pelletier, D.; Nelson, S.J. MR spectroscopic imaging of glutathione in the white and gray matter at 7 T with an application to multiple sclerosis. *Magn. Reson. Imaging* **2010**, *28*, 163–170. [CrossRef] [PubMed]

44. Pocernich, C.B.; Butterfield, D.A. Elevation of glutathione as a therapeutic strategy in Alzheimer disease. *Biochim. Biophys. Acta (BBA) Mol. Basis Dis.* **2012**, *1822*, 625–630. [CrossRef] [PubMed]

45. Lu, S.C. Regulation of glutathione synthesis. *Mol. Asp. Med.* **2009**, *30*, 42–59. [CrossRef]

46. Dalton, T.P.; Chen, Y.; Schneider, S.N.; Nebert, D.W.; Shertzer, H.G. Genetically altered mice to evaluate glutathione homeostasis in health and disease. *Free. Radic. Biol. Med.* **2004**, *37*, 1511–1526. [CrossRef] [PubMed]

47. Ibi, M.; Sawada, H.; Kume, T.; Katsuki, H.; Kaneko, S.; Shimohama, S.; Akaike, A. Depletion of Intracellular Glutathione Increases Susceptibility to Nitric Oxide in Mesencephalic Dopaminergic Neurons. *J. Neurochem.* **2002**, *73*, 1696–1703. [CrossRef]

48. Franco, R.; A Cidlowski, J. Apoptosis and glutathione: Beyond an antioxidant. *Cell Death Differ.* **2009**, *16*, 1303–1314. [CrossRef]

49. Franco, R.; Panayiotidis, M.I.; Cidlowski, J.A. Glutathione Depletion Is Necessary for Apoptosis in Lymphoid Cells Independent of Reactive Oxygen Species Formation. *J. Biol. Chem.* **2007**, *282*, 30452–30465. [CrossRef]

50. Hilton, G.; Nuñez, J.L.; McCarthy, M. Sex differences in response to kainic acid and estradiol in the hippocampus of newborn rats. *Neuroscience* **2003**, *116*, 383–391. [CrossRef]

51. Fu, J.; Xue, R.; Gu, J.; Xiao, Y.; Zhong, H.; Pan, X.; Ran, R. Neuroprotective effect of calcitriol on ischemic/reperfusion injury through the NR3A/CREB pathways in the rat hippocampus. *Mol. Med. Rep.* **2013**, *8*, 1708–1714. [CrossRef]

52. Garcion, E.; Wion-Barbot, N.; Montero-Menei, C.N.; Berger, F.; Wion, D. New clues about vitamin D functions in the nervous system. *Trends Endocrinol. Metab.* **2002**, *13*, 100–105. [CrossRef]

53. Groves, N.J.; McGrath, J.J.; Burne, T.H. Vitamin D as a Neurosteroid Affecting the Developing and Adult Brain. *Annu. Rev. Nutr.* **2014**, *34*, 117–141. [CrossRef] [PubMed]

54. Guo, F.; Yue, H.; Wang, L.; Ding, C.; Wu, L.; Wu, Y.; Gao, F.; Qin, G. Vitamin D supplement ameliorates hippocampal metabolism in diabetic rats. *Biochem. Biophys. Res. Commun.* **2017**, *490*, 239–246. [CrossRef]

55. Ibi, M.; Sawada, H.; Nakanishi, M.; Kume, T.; Katsuki, H.; Kaneko, S.; Shimohama, S.; Akaike, A. Protective effects of 1α,25-(OH)2D3 against the neurotoxicity of glutamate and reactive oxygen species in mesencephalic culture. *Neuropharmacology* **2001**, *40*, 761–771. [CrossRef]

56. Kajta, M.; Makarewicz, D.; Zieminska, E.; Jantas, D.; Domin, H.; Lason, W.; Kutner, A.; Łazarewicz, J.W. Neuroprotection by co-treatment and post-treating with calcitriol following the ischemic and excitotoxic insult in vivo and in vitro. *Neurochem. Int.* **2009**, *55*, 265–274. [CrossRef]

57. Balden, R.; Selvamani, A.; Sohrabji, F. Vitamin D Deficiency Exacerbates Experimental Stroke Injury and Dysregulates Ischemia-Induced Inflammation in Adult Rats. *Endocrinology* **2012**, *153*, 2420–2435. [CrossRef] [PubMed]

58. Brewer, L.D.; Thibault, V.; Chen, K.-C.; Langub, M.C.; Landfield, P.W.; Porter, N.M. Vitamin D Hormone Confers Neuroprotection in Parallel with Downregulation of L-Type Calcium Channel Expression in Hippocampal Neurons. *J. Neurosci.* **2001**, *21*, 98–108. [CrossRef]

59. Daumas, A.; Daubail, B.; Legris, N.; Jacquin-Piques, A.; Sensenbrenner, B.; Denimal, D.; Lemaire-Ewing, S.; Duvillard, L.; Giroud, M.; Béjot, Y. Association between Admission Serum 25-Hydroxyvitamin D Levels and Functional Outcome of Thrombolyzed Stroke Patients. *J. Stroke Cerebrovasc. Dis.* **2016**, *25*, 907–913. [CrossRef] [PubMed]

60. Eyles, D.W.; Smith, S.; Kinobe, R.; Hewison, M.; McGrath, J.J. Distribution of the Vitamin D receptor and 1α-hydroxylase in human brain. *J. Chem. Neuroanat.* **2005**, *29*, 21–30. [CrossRef]

61. De Abreu, D.A.F.; Eyles, D.; Féron, F. Vitamin D, a neuro-immunomodulator: Implications for neurodegenerative and autoimmune diseases. *Psychoneuroendocrinology* **2009**, *34*, S265–S277. [CrossRef]

62. Paintlia, M.K.; Singh, I.; Singh, A.K. Effect of vitamin D3 intake on the onset of disease in a murine model of human Krabbe disease. *J. Neurosci. Res.* **2014**, *93*, 28–42. [CrossRef]

63. Park, K.-Y.; Chung, P.-W.; Kim, Y.B.; Moon, H.-S.; Suh, B.-C.; Won, Y.S.; Kim, J.-M.; Youn, Y.C.; Kwon, O.-S. Serum Vitamin D Status as a Predictor of Prognosis in Patients with Acute Ischemic Stroke. *Cerebrovasc. Dis.* **2015**, *40*, 73–80. [CrossRef]

64. Turetsky, A.; Goddeau, R.P.; Henninger, N. Low Serum Vitamin D Is Independently Associated with Larger Lesion Volumes after Ischemic Stroke. *J. Stroke Cerebrovasc. Dis.* **2015**, *24*, 1555–1563. [CrossRef] [PubMed]

65. Mandell, E.; Seedorf, G.J.; Ryan, S.; Gien, J.; Cramer, S.D.; Abman, S.H. Antenatal endotoxin disrupts lung vitamin D receptor and 25-hydroxyvitamin D 1α-hydroxylase expression in the developing rat. *Am. J. Physiol. Cell. Mol. Physiol.* **2015**, *309*, L1018–L1026. [CrossRef] [PubMed]

66. Mutlu, M.; Sarıaydın, M.; Aslan, Y.; Şebnem, K.; Dereci, S.; Kart, C.; Yaman, S.Ö.; Kural, B. Status of vitamin D, antioxidant enzymes, and antioxidant substances in neonates with neonatal hypoxic-ischemic encephalopathy. *J. Matern. Neonatal Med.* **2015**, *29*, 1–5. [CrossRef]
67. Konety, B.R.; Somogyi, G.; Atan, A.; Muindi, J.; Chancellor, M.B.; Getzenberg, R.H. Evaluation of intraprostatic metabolism of 1,25-dihydroxyvitamin D(3) (calcitriol) using a microdialysis technique. *Urology* **2002**, *59*, 947–952. [CrossRef]
68. LaMonte, G.; Tang, X.; Chen, J.L.-Y.; Wu, J.; Ding, C.-K.C.; Keenan, M.M.; Sangokoya, C.; Kung, H.-N.; Ilkayeva, O.; Boros, L.G.; et al. Acidosis induces reprogramming of cellular metabolism to mitigate oxidative stress. *Cancer Metab.* **2013**, *1*, 23. [CrossRef] [PubMed]
69. Clark, R.S.B.; Empey, P.E.; Bayır, H.; Rosario, B.L.; Poloyac, S.M.; Kochanek, P.M.; Nolin, T.D.; Au, A.K.; Horvat, C.M.; Wisniewski, S.R.; et al. Phase I randomized clinical trial of N-acetylcysteine in combination with an adjuvant probenecid for treatment of severe traumatic brain injury in children. *PLoS ONE* **2017**, *12*, e0180280. [CrossRef]
70. Nemere, I.; Wilson, C.; Jensen, W.; Steinbeck, M.; Rohe, B.; Farach-Carson, M.C. Mechanism of 24,25-dihydroxyvitamin D3-mediated inhibition of rapid, 1,25-dihydroxyvitamin D3-induced responses: Role of reactive oxygen species. *J. Cell. Biochem.* **2006**, *99*, 1572–1581. [CrossRef]
71. Igarashi, H.; Suzuki, Y.; Huber, V.J.; Ida, M.; Nakada, T. N-acetylaspartate Decrease in Acute Stage of Ischemic Stroke: A Perspective from Experimental and Clinical Studies. *Magn. Reson. Med. Sci.* **2015**, *14*, 13–24. [CrossRef]
72. Xu, S.; Waddell, J.; Zhu, W.; Shi, D.; Marshall, A.D.; McKenna, M.C.; Gullapalli, R.P. In vivo longitudinal proton magnetic resonance spectroscopy on neonatal hypoxic-ischemic rat brain injury: Neuroprotective effects of acetyl-L-carnitine. *Magn. Reson. Med.* **2015**, *74*, 1530–1542. [CrossRef] [PubMed]
73. Jiménez-Xarrié, E.; Davila, M.; Candiota, A.P.; Delgado-Mederos, R.; Ortega-Martorell, S.; Julià-Sapé, M.; Arús, C.; Martí-Fàbregas, J. Brain metabolic pattern analysis using a magnetic resonance spectra classification software in experimental stroke. *BMC Neurosci.* **2017**, *18*, 13. [CrossRef] [PubMed]
74. Mlynárik, V.; Kohler, I.; Gambarota, G.; Vaslin, A.; Clarke, P.G.; Gruetter, R. Quantitative proton spectroscopic imaging of the neurochemical profile in rat brain with microliter resolution at ultra-short echo times. *Magn. Reson. Med.* **2007**, *59*, 52–58. [CrossRef] [PubMed]
75. Van Der Zijden, J.P.; Bouts, M.J.R.J.; Wu, O.; Roeling, T.A.P.; Bleys, R.L.; Van Der Toorn, A.; Dijkhuizen, R.M. Manganese-Enhanced MRI of Brain Plasticity in Relation to Functional Recovery after Experimental Stroke. *Br. J. Pharmacol.* **2007**, *28*, 832–840. [CrossRef]
76. Shemesh, N.; Rosenberg, J.T.; Dumez, J.-N.; Muniz, J.A.; Grant, S.C.; Frydman, L. Metabolic properties in stroked rats revealed by relaxation-enhanced magnetic resonance spectroscopy at ultrahigh fields. *Nat. Commun.* **2014**, *5*, 4958. [CrossRef] [PubMed]

Article

NAC and Vitamin D Improve CNS and Plasma Oxidative Stress in Neonatal HIE and Are Associated with Favorable Long-Term Outcomes

Dorothea D Jenkins [1,*], Hunter G Moss [2], Truman R Brown [2], Milad Yazdani [2], Sudhin Thayyil [3], Paolo Montaldo [3], Maximo Vento [4], Julia Kuligowski [4], Carol Wagner [1], Bruce W Hollis [1] and Donald B Wiest [5]

[1] Division of Neonatology, Department of Pediatrics, Medical University of South Carolina, 10 McClennan Banks Drive, Charleston, SC 29425, USA; wagnercl@musc.edu (C.W.); hollisb@musc.edu (B.W.H.)

[2] Center for Biomedical Imaging, Department of Radiology, Medical University of South Carolina, Charleston, SC 29425, USA; mossh@musc.edu (H.G.M.); brotrr@musc.edu (T.R.B.); yazdani@musc.edu (M.Y.)

[3] Centre for Perinatal Neuroscience, Imperial College London, London W12 0HS, UK; s.thayyil@imperial.ac.uk (S.T.); p.montaldo@imperial.ac.uk (P.M.)

[4] Neonatal Research Group, Health Research Institute Hospital La Fe, 46026 Valencia, Spain; maximo.vento@uv.es (M.V.); julia.kuligowski@uv.es (J.K.)

[5] Department of Clinical Pharmacy and Outcomes Sciences, College of Pharmacy, Medical University of South Carolina, Charleston, SC 29425, USA; wiestdb@musc.edu

* Correspondence: jenkd@musc.edu; Tel.: +1-843-792-2112

Citation: Jenkins, D.D.; Moss, H.G.; Brown, T.R.; Yazdani, M.; Thayyil, S.; Montaldo, P.; Vento, M.; Kuligowski, J.; Wagner, C.; Hollis, B.W.; et al. NAC and Vitamin D Improve CNS and Plasma Oxidative Stress in Neonatal HIE and Are Associated with Favorable Long-Term Outcomes. *Antioxidants* 2021, 10, 1344. https://doi.org/10.3390/antiox10091344

Academic Editors: Gaspar Ros Berruezo and Stanley Omaye

Received: 21 July 2021
Accepted: 23 August 2021
Published: 25 August 2021

Publisher's Note: MDPI stays neutral with regard to jurisdictional claims in published maps and institutional affiliations.

Abstract: N-acetylcysteine (NAC) and vitamin D provide effective neuroprotection in animal models of severe or inflammation-sensitized hypoxic ischemic encephalopathy (HIE). To translate these FDA-approved drugs to HIE neonates, we conducted an early phase, open-label trial of 10 days of NAC (25, 40 mg/kg q12h) + 1,25(OH)$_2$D (calcitriol 0.05 mg/kg q12h, 0.03 mg/kg q24h), (NVD), for pharmacokinetic (PK) estimates during therapeutic hypothermia and normothermia. We paired PK samples with pharmacodynamic (PD) targets of plasma isoprostanoids, CNS glutathione (GSH) and total creatine (tCr) by serial MRS in basal ganglia (BG) before and after NVD infusion at five days. Infants had moderate ($n = 14$) or severe HIE ($n = 16$), funisitis (32%), and vitamin D deficiency (75%). NVD resulted in rapid, dose-responsive increases in CNS GSH and tCr that correlated positively with plasma [NAC], inversely with plasma isofurans, and was greater in infants with lower baseline [GSH] and [tCr], suggesting increases in these PD markers were titrated by neural demand. Hypothermia and normothermia altered NAC PK estimates. NVD was well tolerated. Excluding genetic syndromes (2), prolonged ECMO (2), lost-to-follow-up (1) and SIDS death (1), 24 NVD treated HIE infants have no evidence of cerebral palsy, autism or cognitive delay at 24–48 months. These data confirm that low, safe doses of NVD in HIE neonates decreased oxidative stress in plasma and CNS, improved CNS energetics, and are associated with favorable developmental outcomes at two to four years.

Keywords: N-acetylcysteine; vitamin D; neonatal HIE; oxidative stress; MRS

1. Introduction

Therapeutic hypothermia provides significant neuroprotection for half of neonates with uncomplicated hypoxic ischemic encephalopathy (HIE), but not in more severe HIE or HIE complicated by chorioamnionitis [1–6]. Oxidative stress is both an early and persistent contributor to pathology in HIE [7–11]. Glutathione (GSH) is the major intracellular antioxidant which scavenges reactive oxygen species (ROS) and is essential for cell survival [12,13]. Evidence in animal models and humans indicate that oxidative stress depletes reduced GSH within two hours, which is not mitigated by hypothermia alone [14–16]. Increased production of superoxide and lipid peroxides persists for days to weeks after moderate to severe HI/stroke, with significant depletion of GSH in the striatum, hippocampus, cortex, and cerebellum [17–19]. Additionally, the decrease in GSH potentiates glutamate toxicity and FAS-activated cell death [20–23]. Increasing intracellular antioxidant reserves in the brain very early after HI may be key to halting progression of neural cell death and improving neuroprotection for HIE infants who do not respond to hypothermia.

Oxidative stress is a trigger for many inflammatory cascades, and early restoration of normal intracellular redox potential moderates these same pathways. Increasing the CNS pool of available GSH in the metabolically active basal ganglia (BG) may improve neuroprotection after significant neonatal HIE [24]. N-acetylcysteine (NAC) effectively mitigates oxidative stress in animals and humans [9,25,26]. NAC crosses the blood–brain barrier, provides the rate-limiting substrate for GSH synthesis, increases intracellular GSH concentrations and improves cell survival in animal models and humans with multiple etiologies of oxidative stress [9,27–36]. In preclinical work, we previously demonstrated that NAC plus hypothermia improved outcomes over hypothermia alone in neonatal female, but not male, rats subjected to severe HI [37]. When $1,25(OH)_2D$ (calcitriol) was co-administered with NAC and hypothermia, male neonatal rats also showed decreased neuroinflammation and improved neuroprotection [16]. Active vitamin D is a neuro-steroid involved in myelination, neuroplasticity and normal development [38–40]. While anti-inflammatory effects are well described, $1,25(OH)_2D$ also induces synthesis of glutathione reductase, the enzyme responsible for regeneration of GSH from oxidized glutathione disulfide, and thereby increases GSH [41–45]. We postulated that NAC plus $1,25(OH)_2D$ (NVD) would synergistically increase GSH, decrease oxidative stress and improve developmental outcomes in human neonates with severe and complicated HIE [16,46,47].

Therapeutic development of antioxidants in clinical translational neuroscience has been hampered by a lack of careful measurement of biomarkers in target tissue to ensure dosing for pharmacodynamic effect (PD). Biomarkers of oxidative stress include reduced glutathione (GSH) and protein and lipid peroxidation products, which are usually measured in blood. MR spectroscopy (MRS) can directly measure CNS GSH in vivo after injury or treatment [14,29,48–51]. Therefore, we designed this translational study to determine (1) the doses of NAC and $1,25(OH)_2D$ that effectively mitigate oxidative stress in CNS and blood in HIE neonates treated with hypothermia, (2) the duration of NAC and $1,25(OH)_2D$ effect on CNS metabolites and any dose-limiting side effects during hypothermia or normothermia, and (3) the association of improved oxidative stress biomarkers with neurodevelopmental outcomes at >2 years of age.

In a previous rapid communication, we reported our validation of MRS quantification of GSH in this cohort of convalescing HIE infants, with mean BG [GSH] of 1.6 ± 0.2 mM on day of life (DOL) 5 after hypothermia treatment, which is markedly lower than 2.5 ± 0.8 mM previously reported in healthy term neonates [52]. We also reported that NAC with or without calcitriol rapidly and significantly increased [GSH] in the BG to 1.93 ± 0.23 mM ($p < 0.0001$) on DOL 5–6 [15].

In this manuscript, we report the results of the complete clinical trial of NAC and calcitriol (NVD) in neonates with HIE undergoing hypothermia. We quantified plasma lipid peroxidation products and CNS metabolomics with paired plasma NVD concentrations before and after NVD and employ novel, serial MRS to show PD dose response and duration of GSH response in the BG, a major CNS target for neuroprotection in HIE. We show that NAC and calcitriol were safe and determine the dose tolerability and pharmacokinetics (PK) during hypothermia and normothermia. Finally, we report on developmental outcomes of study participants from two to four years of age.

2. Materials and Methods

2.1. Study Design Overview

We enrolled 30 neonates with moderate to severe HIE receiving hypothermia in this Institutional Review Board (IRB)-approved open-label, early phase study (NCT 04643821). We administered intravenous (iv) NAC and calcitriol (NVD) daily for 10 days. We obtained blood samples for NVD PK estimates and oxidative stress markers around the first dose and during 72 h of hypothermia (HT) and around the 10th–11th dose on DOL 5–6 during subsequent normothermia (NT). We paired plasma PK and oxidative stress samples around the 10th or 11th doses with MRS before and after NVD infusion on DOL 5–6, to determine CNS PD dose response in [GSH] and duration of the CNS effect. We were powered to detect 20% [GSH] change of NVD with 20 paired MRS datasets. Our primary safety outcomes were hypotension (NAC) and hypercalcemia (calcitriol). MRS quantification was blinded to dose and condition.

2.1.1. Consent and Enrollment

This study was approved by the Medical University of South Carolina Institutional Review Board. We approached parents within 6 h of birth and obtained written informed consent according to the Declaration of Helsinki, prior to enrolling 30 consecutive neonates with HIE in MUSC's Neonatal Intensive Care Unit from 2014 to 2017. All infants had moderate to severe HIE by modified Sarnat staging [53,54], and qualified for cooling for 72 h according to standard HIE criteria: Gestational age $\geq$ 34 weeks, and they had one factor indicating an acute sentinel event (cord or baby pH $\leq$ 7.0, base deficit $\geq$ -13, Apgar score $\leq$ 5 at 10 min, continued resuscitation after 5 min due to absence of respiratory effort), and two signs of stage 2/3 encephalopathy [55].

2.1.2. Standard of Care: HIE Protocol and Hypothermia Treatment

Infants were cooled to 33–33.5 °C T_r for 72 h (Criticool™, Belmont Medical Technologies, Billerica, MA), and rewarmed at 0.2 °C/h. Infants received nothing by mouth through rewarming other than oral care with breast milk. Parenteral nutrition without cysteine, but with vitamin D_3 400 IU/day and lipids (0.5–2 gm/kg/d), was administered during hypothermia. Feeds were introduced if clinically stable on DOL 4–5. Infants received IV morphine 0.02 mg/kg every 4 h for comfort during hypothermia that was increased as clinically required in infants with pulmonary hypertension (PPHN). Standard monitoring during hypothermia included serum lactate, ionized and total calcium, electrolytes, clotting studies, liver function tests, cardiac and renal panels, and circulating cell counts. Continuous 20 montage video EEG was monitored from admission until completion of rewarming. Seizures were defined as electrographic seizure activity. Head ultrasound with Doppler blood flow of three cerebral arteries was obtained within 12 h of admission using a Philips Epic 5 scanner (Koninklijke Philips N.V., Amsterdam, The Netherlands) with a 9–4 MHz curvilinear transducer, and we recorded the average resistive index and time average maximum velocity. Echocardiograms were obtained in all infants. PPHN was defined as evidence of right to left intra- or extracardiac shunting through the patent ductus arteriosus. An MRI and MRS were obtained on day of life 5–6 after rewarming, on a 3T Siemens Skyra, using standard sequences for T1, T2, ADC, and DKI. For MRS, single voxels were placed in the left BG and right frontal white matter. MRI scans were read by a single neuroradiologist (MY).

2.1.3. NAC and Calcitriol Infusion

The study drugs were administered by IV within 4–9 h after birth and continued until DOL 10 or discharge. NAC was infused over 60 min, and calcitriol by IV push. For the first 20 participants the NAC dose was held constant at 25 mg/kg/dose every 12 h while determining the optimal dose and timing of calcitriol. For the next 10 participants, the NAC dose was increased to 40 mg/kg, keeping the calcitriol dose constant. Treatment groups were NAC 25 mg/kg/dose + calcitriol 0.05 mcg/kg/dose iv q12h (n = 10), NAC 25 mg/kg/dose q12h + calcitriol 0.03 mcg/kg/dose q24h (n = 10), or NAC 40 mg/kg/dose q12h + calcitriol 0.03 mcg/kg/dose q24h (n = 10). Both study drugs were held if an infant had untreated/refractory hypotension at time of dosing (mean BP < 40 mmHg). Calcitriol was held for ionized calcium (iCa^{++}) > 1.3 mmol/L, and we limited calcium in parenteral nutrition. An early calcitriol PK analysis estimated serum calcitriol half-life to be 28.2 h during hypothermia and mean peak [1,25(OH)$_2$D] = 263 $\pm$ 72 pmol/L (first dose), increasing to 434 $\pm$ 146 pmol/L (sixth dose), indicating continued accumulation with 0.05 mcg/kg/dose q12h. Subsequently the calcitriol dose was decreased to 0.03 mcg/kg/dose q24h (n = 20), but still within the range of effective doses in our preclinical studies [16].

2.2. Magnetic Resonance Spectroscopy

2.2.1. MRS Protocol

On DOL 5–6 after rewarming, clinically stable HIE neonates had MRS scans with paired blood samples for NVD and lipid peroxidation products before and after the 10th or 11th NVD dose, using a single voxel placed in either the BG or frontal white matter, with stimulated echo acquisition mode (STEAM), echo time (TE) 20 ms, and point resolved spectroscopy (PRESS) TE 270 ms on a Siemens 3T Skyra, as previously described [15]. Infants received both NAC and calcitriol, or only NAC if the calcitriol dose was held or not given at that time with 24 h dosing, as scans had to be conducted at night. The neuroimaging protocol (Figure 1) consisted of: (1) first MRS immediately preceding NVD dose administration (trough); (2) calcitriol and NAC infusion during the clinical MRI; (3) second MRS immediately after clinical MRI (12–30 min after NVD infusion, peak); (4) third delayed MRS (2–6 h after infusion as the clinical schedule allowed, post-peak) to estimate the duration of NAC effect on [GSH] in the CNS; (5) MRS/diffusion imaging obtained between 10–40 d (convalescent). Sedation with low dose morphine or lorazepam was administered for clinical indications during acute scans. Scans required 90 min for the acute protocol, 12 min for the delayed scan. The PI (neonatologist) and neonatal intensive care unit nurse were present during all scans.

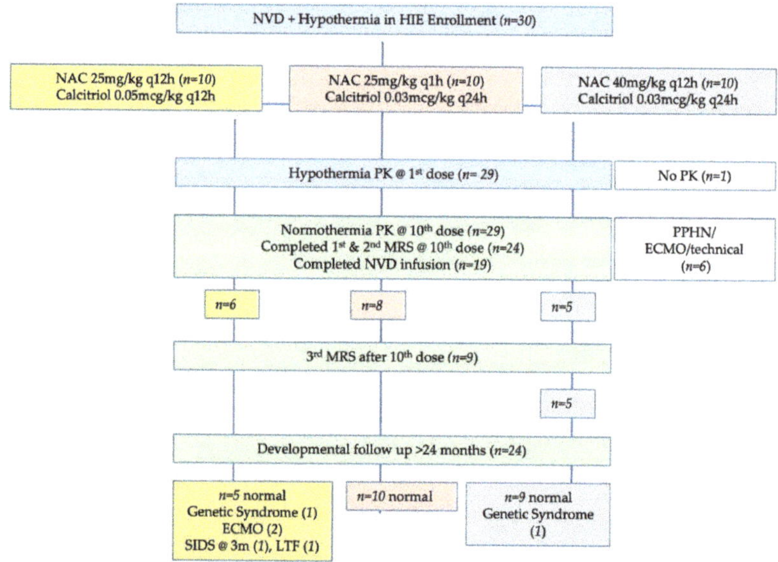

Figure 1. Study flow diagram. Note that the number of infants with developmental follow-up flows from the original number of infants in each dosing group, not just the infants with complete NVD infusions or complete MRS data. The number of infants with useable data for each MRS metabolite may vary slightly from these numbers due to MRS quality controls for spectral fit which are unique to each metabolite.

2.2.2. LCModel Data Fitting

MRS spectra were fit with LCModel by an investigator blinded to subject and dose, using a custom basis set that included GSH with a cysteinyl peak at 2.95 ppm, as well as standard metabolites: total Choline (tCho: GPC+PCho), total creatine (tCr: Cr+PCr), total N-acetylaspartate (NAA: NAA+NAAG), myoinositol (mIns), glutamate and glutamine (GLX)] [15]. Absolute metabolite concentrations were determined with water normalization from the unsuppressed water scan. Representative 3T MRS of the GSH peak at 2.95 ppm and validation of quantification has been previously published [15]. We excluded spectra for low signal/noise < 5, or coefficient of variation > 15% by LCModel fit.

2.3. Pharmacokinetic Studies

2.3.1. Drug Concentrations

Blood samples were collected for HT and NT PK: before and 0.5 h, 1 h, 11.5 h after first (HT) and tenth (NT) NAC/NVD doses, and for steady-state trough concentration (Cmin$_{ss}$) prior to sixth (HT) and thirteenth doses (NT). Blood was collected in sodium EDTA (NAC) and plain tubes (vitamin D), and plasma/serum was aliquoted and frozen at $-80\ °C$ until assay. CSF obtained at 24 h with the peak NVD blood sample was frozen at $-80\ °C$ until assay. Total plasma NAC concentrations were determined using a modified, sensitive and specific HPLC method with a limit of detection of 0.15 mmol/L [56]. Serum 25(OH)D & 1,25(OH)$_2$D concentrations were measured for 23 infants by RIA with detection limits of 2.5 mol/L for 25(OH)D and 36 pmol/L for 1,25(OH)$_2$D [57]. Frozen aliquots of serum for 20 infants were shipped on dry ice for analysis of isoprostanoids markers of oxidative stress by LC-MS, as previously reported [18].

2.3.2. PK Analyses

Non-compartmental modeling methods (PK Solutions 2.0; Summit Research Services, Montrose, Colorado) generated PK parameters for NAC and 1,25(OH)$_2$D: terminal elimination half-life ($t_{1/2}$), volume of distribution (Vd), total body clearance (CL) and area under the concentration curve (AUC). Statistical comparisons between doses or HT and NT were conducted using a paired or unpaired 2-sided Student t test, with significant p-value < 0.05. (Graphpad Software 5.00, San Diego, CA, USA).

2.3.3. Developmental Follow-Up Testing

We performed standard developmental testing in high-risk follow-up or developmental clinics for most participants between 18–36 months of age. This battery of tests includes the Peabody Developmental Motor Scales, which have a high correlation with the Bayley III motor scores (r = 0.79–0.88 for fine, and gross/total motor scores, respectively [58]; Cognitive adaptive test (CAT) and Clinical linguistics and auditory milestone scale (CLAMS), which are both sensitive and specific for cognitive delays defined as the Bayley cognitive scaled score < 70 [59] and the Modified Checklist for Autism in Toddlers (MCHAT). Parental reports of development by phone were also obtained at 4 years of age to discern any intervening issues with language or speech. Two families moved out of the area and only provided parental reports of development and the use of special services.

2.3.4. Statistical Analysis

For 24 subjects with CNS metabolite concentrations regardless of completion of NVD infusion (intent-to-treat), two-tailed paired t-tests compared changes in metabolite concentrations in HIE neonates pre- and post-infusion with significance $p < 0.05$ (IBM SPSS® v.25, Armonk, NY, USA). ANOVAs were used to compare repeated measures of non-linear MRS metabolite concentrations for three time points within a subset of 10 subjects, with pairwise comparisons and Bonferroni correction. Partial eta squared values are reported for attributing the variance in metabolite concentration to time of scan around NVD infusion. Independent variables of NAC dose and sex were included in analyses. We performed both intent-to-treat analyses of CNS metabolites and efficacy analysis (infants who received full dosing) to account for infusion problems and lack of reliable dose delivery during some MR scans. Pearson's or Spearman's correlation coefficients were generated for comparisons of plasma NAC and lipid peroxidation products, and CNS metabolite concentrations, as appropriate.

3. Results

3.1. Clinical Characteristics

We enrolled 30 neonates with moderate to severe HIE receiving hypothermia in this IRB-approved open-label, early phase study (Figure 1). Study drugs were administered within 4–9 h after birth and continued through 10 days of life through a central or peripheral IV line.

Clinical demographic data of the neonatal HIE cohort are presented in Table 1. There were twice as many males as females. Infants exhibited neurologic signs of moderate to severe encephalopathy prior to enrollment and were hypothermic at first study drug administration. Out of the 19 placentas sent to pathology we demonstrated acute chorioamnionitis with funisitis in one third of HIE infants. One placenta showed fetal vasculopathy with thrombosis and many avascular villi. The numbers of infants in the worst quartiles for pH (6.5–6.8) and base deficit (>−20) are also given.

Table 1. Demographics of HIE infants. Mean (SD) or number of subjects with specific conditions.

	N = 30
Gestational age (weeks), mean (SD)	38.2 ± 1.5
Birth weight (grams), mean (SD)	3248 ± 545
Males/Females	20/10
pH of cord or neonatal blood, median (IQR)	6.91 (6.82, 6.99)
Cord pH 6.5–6.8, number of subjects	6
Base deficit of cord or neonatal blood, median (IQR)	−18.0 (−20, −13.6)
Base deficit $\geq$ −20, number of subjects	10
Serum lactate (mM) < 6 h after birth, mean (SD)	8.9 ± 4.6
Apgar score at 1 min, mean (SD)	1.4 ± 1.3
Apgar score at 5 min, mean (SD)	3.2 ± 2.0
Apgar score at 10 min, mean (SD)	5.4 ± 2.1
Chest compressions, number of subjects	10
HIE stage 2/3, number of subjects	14/16
Placental histology: Chorioamnionitis	8/19 (42%)
Funisitis	6/19 (32%)
Clinical Sepsis/Pneumonia	8
Complete abruption or feto-maternal hemorrhage	5
Seizures	7
PPHN	6
ECMO	2
Congenital syndrome	2
MRI abnormalities, total number of patients	13
Minor T1 & T2 signal abnormalities	10
Basal ganglia	1
Periventricular white matter	7
Cortical/subinsular	1/1
Hemorrhages	
Periventricular/Cortical/Occipital/Cerebellum	1/1/1/1
Perinatal arterial stroke	1
Gr 1–2 IVH	4

Six infants had PPHN prior to enrollment and four were treated with inhaled nitric oxide. Two of these infants required prolonged and complicated courses of extracorporeal membrane oxygenation (ECMO): one was placed on veno-venous by-pass and deteriorated over the next 24 h requiring conversion to veno-arterial bypass with overwhelming sepsis and pneumonia. Study drugs were held for 41 h during these events. The second infant was placed on ECMO within a few hours of birth and did not receive the first dose of study drug until blood pressure was stable on bypass. This infant later suffered an emergent circuit change due to circuit thrombosis with subsequent major pulmonary and CNS hemorrhages with hypoxia (1). Two genetic disorders were diagnosed after treatment completion (Prader Willi and congenital myasthenia gravis).

Resistive indices on Doppler ultrasound ($n = 28$) within 24 h of HIE birth were mean RI = 0.72 ± 0.11(0.53–0.94) in the anterior cerebral artery, 0.73 ± 0.11 (0.53–0.90) in the middle cerebral artery, and 0.74 ± 0.10 (0.54–1.0) in the basilar artery. Abnormal RI (<0.65 or >0.8) were noted in 13 infants in the middle cerebral and basilar arteries and 15 infants in the anterior cerebral artery [60,61].

MRI brain parenchymal signal abnormalities were present in 10 infants, largely demonstrating small foci of periventricular signal abnormality characterized by subcentimeter foci of T1 and T2 shortening with associated restricted diffusion (Table 1). Other imaging abnormalities included one patient with right sided subcentimeter cortical hemorrhages, one patient with left subinsular white matter injury, one patient with subcentimeter focus of T2 hyperintensity in the right basal ganglia, one patient with a subcentimeter right cerebellar hemorrhage, and one patient with left middle cerebral artery and additional multifocal small bilateral cortical infarctions. Among the 10 infants with minor abnormal parenchymal findings, four had small volume intraventricular hemorrhage. Among the patients with no brain parenchymal signal abnormalities, one patient had trace intraventricular hemorrhage. Seventeen infants had no identifiable abnormalities on their conventional MRIs.

3.2. NVD Safety and Tolerability

NAC 25 and 40 mg/kg/dose q12h were well tolerated without serious adverse events. We held NAC and calcitriol doses for urgent cannulation for ECMO per IRB-approved protocol. Patients on ECMO received an NAC loading dose to account for increased Vd of the ECMO circuit. Calcitriol dose was not altered.

We held 26 calcitriol doses for elevated iCa^{++} (mean iCa^{++} = 1.42 ± 0.06 mM), which were not related to dose or interval. iCa^{++} was elevated in five out of ten infants receiving calcitriol 0.05 mcg/kg q12h (1.46 ± 0.02 mM), and 18 out of 20 infants receiving calcitriol 0.03 mcg/kg q24h (iCa^{++} 1.41 ± 0.07 mM). Many of the infants had calcitriol doses held while on full feeds, when ionized calcium concentrations generally increased. Therefore, the mean duration of calcitriol dosing was 5.7 ± 2.1 days. We observed no clinical effects of mildly elevated ionized calcium, which was managed by withholding calcitriol dosing and/or decreasing calcium in parenteral nutrition. Elevated iCa^{++} were not associated with elevated total serum calcium concentrations (all < 12 mg/dL).

3.3. NAC and Vitamin D Pharmacokinetics

3.3.1. NAC PK

We found significant differences in NAC PK parameters between HT and NT periods, with lower clearance and longer half-life during hypothermia (Table 2, Supplementary Table S1, Figure 2A–D). Steady-state NAC trough (Cminss) concentrations were low during both periods with q12h dosing. NAC PK did not significantly differ between calcitriol dosing groups or when ECMO patients were excluded from analysis.

Table 2. NAC and 1,25(OH)2D PK during HT and NT.

NAC PK	GA wks	Birth Wt gm	$t_{1/2}$ (hrs) HT vs. NT		Vd (L/kg) HT vs. NT		CL (mL/hr/kg) HT vs. NT		$Cmin_{ss}$ µMol HT vs. NT		$Cmax_{ss}$ µMol HT vs. NT		AUC 0-t HT vs. NT	
Mean NAC 25 mg/kg (SD)	38.1 (1.6)	3210 (456)	5.8 (1.9)	4.2 (1.1)	0.63 (0.29)	0.73 (0.31)	68 (23)	113 (35)	70.8 (54.8)	25.6 (23.6)	365.0 (93.2)	251.1 (84.0)	300.4 (71.8)	200.1 (69.6)
p value HT vs. NT NAC 25 mg/kg			0.0050 *		0.1415 *		0.0006 *		0.0017 *		0.0030 *		0.0018 *	
Mean NAC 40 mg/kg (SD)	38.8 (1.3)	3110 (1137)	4.9 (1.3)	4.0 (1.3)	0.49 (0.13)	0.70 (0.1)	70 (16)	133 (41)	105.8 (49.1)	47.9 (36.8)	585.9 (117.8)	365.6 (71.2)	467 (96.6)	265.7 (71.7)
p value HT vs. NT NAC 40 mg/kg			0.15 *		0.0026 *		0.0018 *		0.0208 *		0.0019 *		0.0021 *	
p value NAC 25 vs. 40 mg/kg/dose											0.12	0.07	**0.0018**	**0.0018**
1,25(OH)$_2$ D PK	GA wks	Birth Wt gm	$t_{1/2}$ (hrs) HT vs. NT		Vd (L/kg) HT vs. NT		CL (mL/hr/kg) HT vs. NT		$Cmin_{ss}$ pMol HT vs. NT		$Cmax_{ss}$ pMol HT vs. NT		AUC 0-t HT vs. NT	
Mean 1,25(OH)$_2$ D 0.05 mcg/kg	37.9 (1.4)	3197 (466)	26.2 (16.0)	18.3 (7.0)	0.65 (0.25)	0.43 (0.45)	24.4 (15.5)	14.1 (11.8)	458 (242)	737 (742)	648 (238)	1330 (1445)		

* Indicate comparisons of HT and NT PK parameters within dosing group.

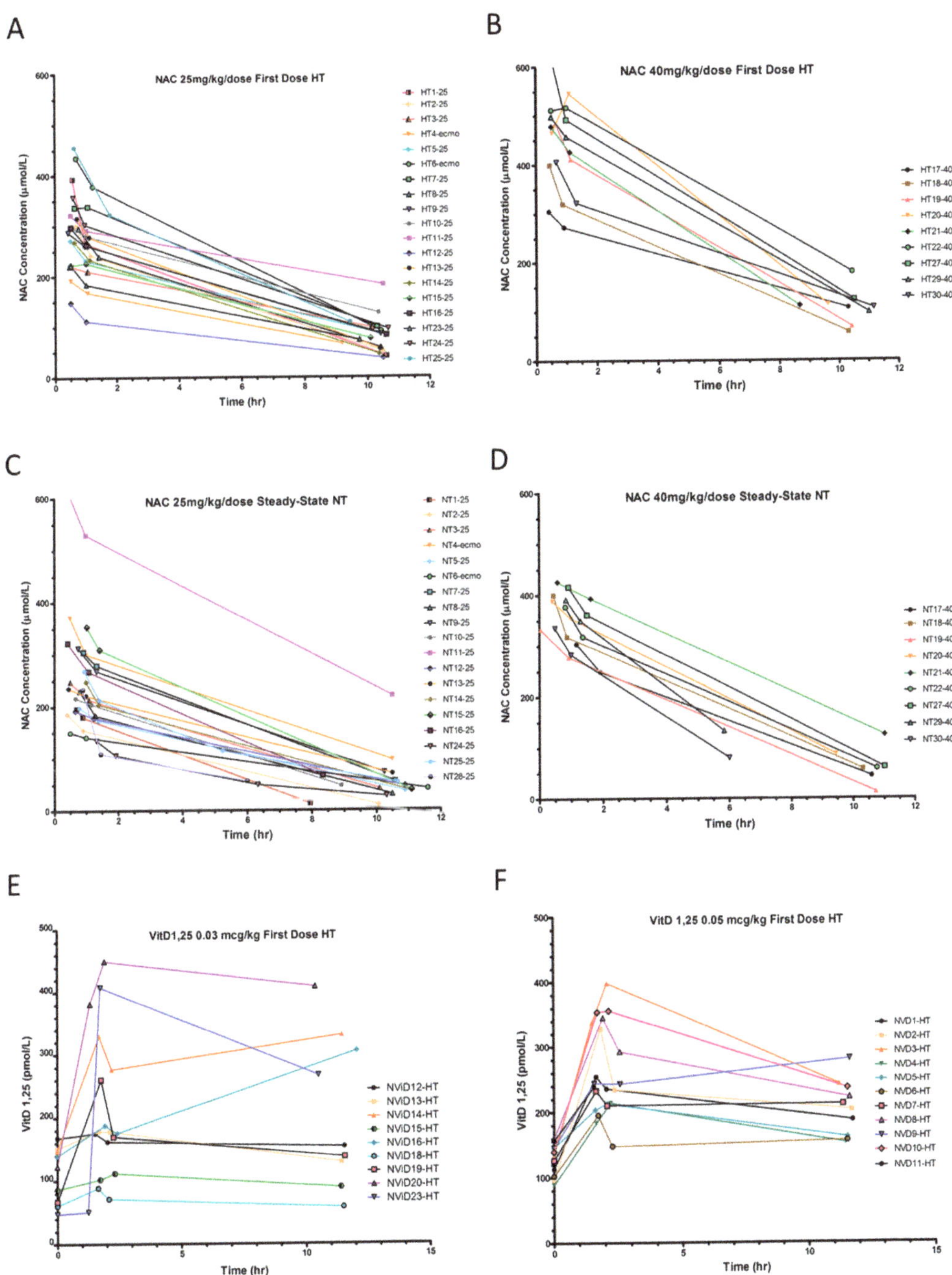

Figure 2. Plasma NAC and serum 1,25(OH)$_2$D concentrations and pharmacokinetics. (**A,B**) Plasma NAC concentrations during hypothermia (HT) and (**C,D**) normothermia (NT), with NAC (**A,C**) 25 mg/kg/dose and (**B,D**) 40 mg/kg/dose. (**E,F**) 1,25(OH)$_2$D serum concentrations during HT with calcitriol (**E**) 0.03 mcg/kg/dose, (**F**) 0.05 mcg/kg/dose.

3.3.2. 25(OH)D and 1,25(OH)$_2$D Serum Concentrations at Birth and Pharmacokinetics

At birth, 78% of 23 HIE neonates sampled were vitamin D insufficient [25(OH)D < 75 nmol/L], 61% were vitamin D deficient [<50 nmol/L], and 30% were severely deficient [<25 nmol/L], similar to our previous HIE cohort [57]. There were significant differences in 25(OH)D serum concentrations at birth by race but not by sex. Black and Hispanic infants had mean 25(OH)D 25 ± 10 nmol/L vs. white infants 62.5 ± 25 nmol/L (p = 0.00025). Furthermore, all black/Hispanic infants were 25(OH)D deficient at birth, range (12.5–45 nmol/L). Among white infants, 31% were deficient, 31% were insufficient, and 38% were 25(OH)D sufficient. 25(OH)D serum concentrations did not increase with 400IU vitamin D$_3$ added to daily parenteral nutrition or enteral feeds.

Mean 1,25(OH)$_2$D serum concentrations were 48 ± 16 pg/mL shortly after birth and did not vary by race or sex. Serum [1,25(OH)$_2$D] increased with the first calcitriol dose by 80.6 ± 25.2 pmol/L (0.03 mcg/kg) and 143.5 ± 62.2 pmol/L (0.05 mcg/kg), but show considerable variability in PK, particularly with the 0.03 mcg/dose (Figure 2E,F). Many patients had concentrations at 60 min that were higher than "peak" concentrations at 30 min, precluding PK estimate calculations. This may have been due to the small dose volume being administered, continued release from vitamin D binding protein [62], or a prolonged equilibration phase. Calcitriol was frequently held during normothermia for elevated iCa^{++}. Therefore, we report calcitriol PK for only six patients receiving 0.05 mcg/kg during HT (Figure 2F, Table S2), which prevented any meaningful comparisons between doses, or HT and NT periods.

3.3.3. NAC and Vitamin D concentrations in CSF

We measured NAC concentration in the CSF in five infants (mean 17.2 ± 23.9 mmol/L, range: 6.1–61.3 mmol/L) within 7–25 min after the second NAC dose (Supplementary Table S1). Compared with the plasma obtained at the same time as CSF, secretion of NAC into CSF was between 1–21% of peak plasma [NAC]. 1,25(OH)$_2$D was not detectable in any CSF sample.

3.4. MRS Metabolite Changes with NVD Infusion

NVD was infused during the routine MRI on DOL 5–6, and MRS scans were performed at 0 h before infusion, at 0.5 h, and 2–6 h after infusion to study the PD time course of [GSH] change in CNS. We obtained adequate MRS spectra and metabolite data in 23–24 neonates at two time points depending on metabolites, and in 10 patients at three time points. For efficacy analyses we excluded MRS if there were problems with the IV or infusion pump resulting in incomplete infusion of NVD prior to second scan (n = 5), and significant delay outside of dosing windows by emergent clinical cases (n = 1). As previously reported, three infants had no NAC infused prior to the second scan and had no increase in [GSH] [15]. One infant had left parietal perinatal stroke, and a voxel was placed in the diffusion-restricted stroke area in addition to the BG.

3.4.1. Dose Responsive Increase in [GSH] with NVD Infusion in Basal Ganglia

For the 10 infants with three scans, [GSH] showed a significant increase acutely and over time (F = 8.2, p = 0.012, partial eta sq = 0.67). Seven out of 10 infants maintained improved [GSH] in BG for four to six hours (Figure 3A). We observed a CNS dose response with a significantly greater increase in BG [GSH] with NAC 40 mg/kg (+32%) versus 25 mg/kg (+18%, p < 0.05, efficacy analysis, Figure 3B). Importantly, peak [GSH] in BG did not exceed the normal range in healthy term neonates with either NAC dose [52]. We previously reported CNS [GSH] in the BG increased significantly 12–30 min after NAC/NVD (p < 0.0001, n = 23, all doses, intent-to-treat, paired t-test), which was not significantly different between NAC alone (n = 9) vs. NVD (n = 14) [15]. Coefficients of variance were ≤8% for [GSH] with LCModel processing of STEAM sequences at TE20ms on Siemens 3T MRI.

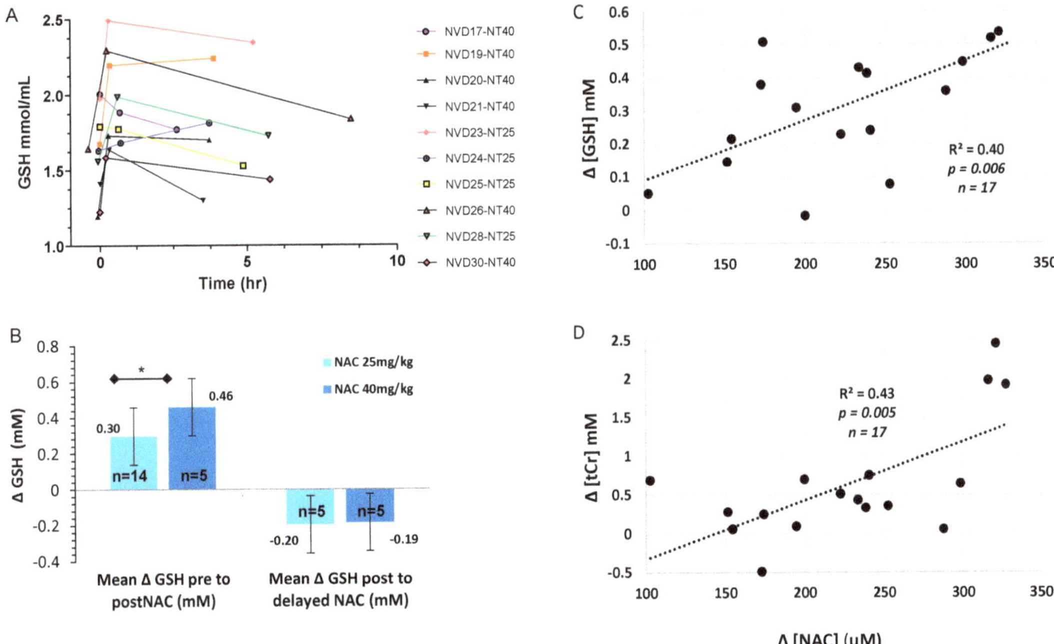

Figure 3. CNS GSH changes over time and by NAC dose. (**A**) [GSH] in BG of the 10 HIE infants with serial MRS on DOL 5–6 immediately before (trough = 0 h), and after NVD (peak = 12–30 min) and delayed (post-peak = 2–6 h), with NAC alone (*n* = 5) or NVD infusion (*n* = 5). (**B**) [GSH] change with NAC 25 mg/kg/dose or 40 mg/kg/dose from trough to peak [GSH] or peak to post-peak (2–6 h) after NAC/NVD infusion (* *p* < 0.05). We included all patients who had the entire dose reliably delivered prior to second scan (efficacy analysis). (**C,D**) Trough to peak increase in plasma [NAC] plotted with the simultaneous increase in basal ganglia metabolites (**C**) [GSH] and (**D**) [tCR] for the 17 subjects with adequate MRS and [NAC] plasma data.

3.4.2. Total Creatine and Choline Change Acutely with NVD Dosing

NVD (all doses) significantly increased tCr, tCho, and mIns within 12–30 min of infusion by paired *t*-test (*n* = 23–24 depending on metabolite, intent to treat, Table 3). Mean NAA increased from pre- to immediately post-NVD (*n* = 24, intent to treat, *p* = 0.051, paired *t*-test), which was significant when infants with infusion problems were excluded (*n* = 18, *p* = 0.042). There was a trend to a dose response for [tCr] and [tCho] in the BG, with greater increase with NAC 40 mg/kg versus 25 mg/kg (both *p* = 0.07).

Table 3. MRS metabolite concentrations for all subjects, Mean (SD).

CNS metabolite (mM)	Acute MRS (DOL 5)			Convalescent MRS
	Pre-NVD Trough (*n* = 24)	Post-NVD Peak (*n* = 24)	Post-NVD Post-Peak (*n* = 10)	DOL 11–40 (*n* = 18)
GSH	1.61 ± 0.28 *	1.93 ± 0.31 *	1.77 ± 0.32	2.05 ± 0.37
Total Creatine	6.49 ± 0.7 *	7.06 ± 0.53 *,†	6.98 ± 0.81 †	6.93 ± 0.55
Total Choline	2.43 ± 0.3 ‡	2.67 ± 0.3 *,‡	2.46 ± 0.30 *	2.63 ± 0.33
NAA	4.62 ± 0.5	4.84 ± 0.46	4.55 ± 0.71	5.11 ± 0.81
Glutamate + glutamine	6.59 ± 0.94	6.37 ± 0.57	7.04 ± 0.74	7.35 ± 1.08
MyoInositol	7.12 ± 0.95 †	7.65 ± 0.91 †	7.13 ± 0.70	7.83 ± 0.97

Comparing MRS metabolites over 3 scans during the acute HIE phase, before and after NVD infusion on DOL5, intent to treat analysis: overall ANOVAs are significant for GSH (*p* = 0.012), tCr (*p* = 0.005), tCho (*p* = 0.001); mIns (*p* = 0.009). Specific groups that were significantly different by pairwise comparisons are noted by * *p* < 0.05, † *p* < 0.005, ‡ *p* < 0.0005.

For the 10 subjects with three scans, other metabolite concentrations were significantly different over the three scans: tCr (F = 10.9, p = 0.005, partial eta sq = 0.73); tCho (F = 20.9, p = 0.001, partial eta sq = 0.83); mIns (F = 9.1, p = 0.009, partial eta sq = 0.7). There were no significant differences in [NAA] or [GLX] within these 10 subjects by repeated measures ANOVA.

3.4.3. Sex Differences in CNS Metabolites

[GSH] in BG before NVD infusion showed a strong tendency to be lower in HIE males (1.59 ± 0.19 mM, n = 17) than in females (1.74 ± 0.18 mM, n = 6, p = 0.06), but [GSH] increase with NVD was not different by sex. There were no sex differences in mean metabolite concentrations for other metabolites, but there were too few females for meaningful repeated measures comparisons.

3.5. Plasma NAC and Lipid Peroxides Correlate with CNS Markers of Oxidative Stress and Energetics

NAC plasma concentrations strongly correlated with the simultaneous increase in CNS [GSH] and [tCr] from trough to peak, before and after the NVD infusion (Figure 3C,D). The absolute change in [ΔGSH] also positively correlated with the absolute change in [ΔtCr] from trough to peak (Figure 4A), demonstrating that CNS energetics improved in conjunction with antioxidant concentrations. Moreover, the lower the baseline [GSH] and [tCr] before dosing (trough), the greater the increase with NAC/NVD (Figure 4B,C). Conversely, the higher the trough [GSH] and [tCr] in BG before NAC/NVD, the less the absolute increase in [GSH] and [tCr].

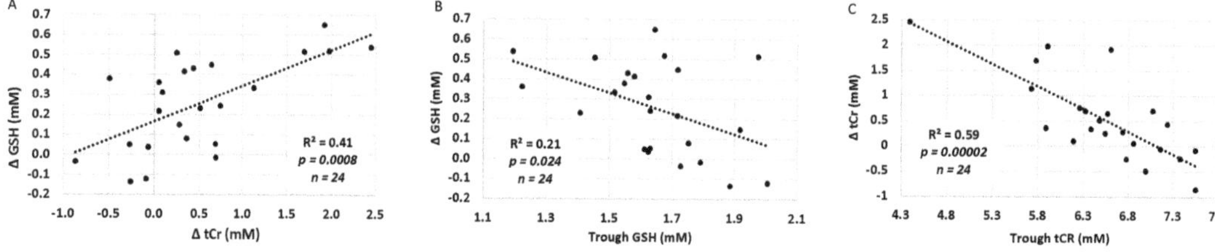

Figure 4. Comparison of GSH and tCr in BG (intent to treat analysis). (**A**) The increase in [GSH] is plotted against the change in [tCR] trough to peak, across all infants with NAC/NVD infusion, including those with only partial infusions. (**B,C**) The increase in (**B**) [GSH] and (**C**) [tCr] are strongly dependent upon the trough concentrations in the BG. There was a significantly greater increase in [GSH] and [tCR] in those with lower CNS GSH and tCr on DOL 5 before dosing (trough).

We also investigated whether plasma concentrations of lipid peroxidation markers, isoprostanoids, were associated with CNS oxidative stress and acutely changed with NAC/NVD infusion in paired plasma samples around DOL 5 MRS. The decrease in Isofuran concentrations from before to immediately after NAC/NVD infusion correlated with simultaneous increase in CNS [GSH], from trough to 30 min peak (r = −0.76, p = 0.006, n = 11, Figure 5A). Several plasma oxidative stress markers were inversely correlated with CNS [GSH] post NVD. Higher CNS [GSH] correlated with lower plasma Neurofurans at one hour post NVD (r_s = −0.88, p = 0.0038, n = 10, Figure 5B); Other correlations were limited due to small numbers of samples but include 5IPF2a-VI concentrations at 30 min after NVD dosing (r = −0.78, p = 0.0022, n = 8). Neither 25- or 1,25(OH)$_2$ plasma concentrations correlated with plasma iso- or neurofurans.

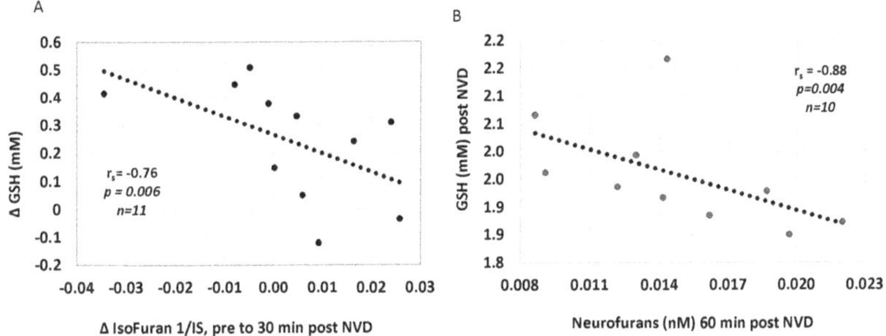

Figure 5. Plasma oxidative stress markers Isofurans and Neurofurans correlate with CNS [GSH] after NVD. (**A**) The increase in [GSH] is strongly correlated with a decrease in Isofurans from pre to post NVD dosing. (**B**) [GSH] after NVD infusion is strongly negatively correlated with Neurofurans at 60 min post infusion. (Spearman's rank correlation).

3.6. Death and Neurodevelopmental Outcomes

The mean days to discharge for 26 HIE infants (excluding prolonged ECMO or genetic syndromes) was 16 ± 12 d. There were no deaths during initial hospitalization. One infant suffered a roll-over accidental death at three months during co-bedding. The two infants with a history of ECMO and one infant with a genetic syndrome were the only participants to require gastrostomy tubes for feeding. The ECMO and genetic syndrome patients had abnormal development with spastic diplegic cerebral palsy (1), global developmental delay (4), and autism (1). One infant was lost to follow-up.

The remaining 24 infants with HIE treated with NVD and hypothermia show normal gross motor, fine motor, and cognitive development at 24–48 months of age, without evidence of cerebral palsy or autism. These 24 infants include 10 infants either exposed to chorioamnionitis or diagnosed with clinical sepsis and pneumonia. Twenty-two had formal developmental assessments at developmental follow-up clinics: the mean Peabody gross motor developmental quotient was 102 ± 9.5, the Cognitive adaptive test (CAT) was 95 ± 9.3, the Clinical linguistics and auditory milestone scale (CLAMS) was 98.9 ± 11.1, and the mean MCHAT was 0.4 ± 0.5, at 33 ± 14 months of age. In the severe HIE group ($n = 13$), the Peabody gross motor developmental quotient was 103 ± 8, CAT 101 ± 10, CLAMs 95 ± 11, MCHAT 0.14 ± 0.4 at 29 ± 12 months of age. Three male infants were receiving therapy for speech delay, and one made significant improvement after myringotomy tube placement. Parents of two infants who moved out of the area reported normal development and no special services.

4. Discussion

In neonates with moderate and severe HIE, oxidative stress continues for days after completion of therapeutic hypothermia, as reflected in low concentrations of reduced glutathione in the basal ganglia before NVD dosing on DOL 5–6. NAC crossed the blood–brain barrier quickly and significantly increased [GSH] in the BG of HIE neonates on DOL 5–6, in a dose responsive manner, with a measurable effect for at least 6 h, while the serum NAC half-life during the same normothermic period was 4.5 h. Both NAC doses resulted in GSH increases, but the 40 mg/kg dose resulted in a 32% increase within 30 min of infusion. The 40 mg/kg dose of NAC appeared safe both systemically and in CNS, as we did not observe evidence of reductive stress [63], and GSH never exceeded reported normal values in term infants.

At the same time, NAC/NVD increased total creatine in the BG, which is strongly correlated with the increase in [GSH]. The simultaneous rapid improvement in tCr indicates that along with restoring CNS GSH reserves, NAC and or NVD improve mitochondrial and cellular creatine kinase and synthetic function [64] and increase important energy substrates in a vulnerable brain region. Further, the increases in GSH and tCr were greater in patients with lower total creatine and greater oxidative stress, and not as great in infants with higher GSH or tCr concentrations. Taken together, these data strongly suggest that NAC 25–40 mg/kg/dose and calcitriol result in rapid synthesis of GSH and increases in CNS energetics that are titrated based on level of CNS demand from oxidative stress in the BG.

This finding is immensely important for the safe clinical translation of these therapeutics. Under normal conditions, the production of intracellular GSH is tightly regulated by availability of the rate-limiting precursor cysteine and direct feedback inhibition of the major synthetic enzyme, glutamate-cysteine ligase [65]. In hypoxic-ischemic and reperfusion conditions, the feedback inhibition via phosphorylation may be impaired, leading to excess GSH and reductive stress even with low doses of NAC and vitamin D. Our MRS data on DOL 5 show that feedback inhibition is intact and there is tight regulation of GSH synthesis to a normal intracellular concentration. Therefore, infants with the greatest oxidative stress can benefit from NVD, without exposing less affected infants to harmful side effects.

Plasma isofurans, which are stable and peak more slowly than isoprostanes [66], were also reduced after NAC/NVD infusion on day five, which strongly correlated with the simultaneous increase in CNS [GSH]. The combined data on CNS and plasma oxidative stress markers indicate that infants are still not metabolically stable during the subacute phase of HIE despite hypothermia treatment, and that NAC rapidly corrects both systemic and CNS oxidative stress and metabolomics. Even infants with severe HIE by neurologic exam, cord pH ≤ 6.8, and MRI lesions seemed to benefit, as motor and cognitive outcomes for this group were within the normal range. The fact that significant structural MRI lesions were found on DOL 5 in one perinatal arterial stroke patient and yet functional outcomes at 24 months of age were favorable without hemiparesis, suggests that this metabolic recovery is very important to functional outcome.

NAC and vitamin D pharmacokinetics differ by hypothermic and normothermic phases of recovery. Similar to findings of other PK trials in neonatal therapeutic hypothermia [67,68], we found longer $t_{1/2}$ and decreased CL of NAC during hypothermia when compared to normothermia. Based on this data, NAC dose and dosing interval need to be adjusted to increase NAC trough steady state concentrations during both HT and NT. NAC 30 mg/kg every eight hours and 40 mg/kg every eight hours during HT and NT, respectively, will maintain troughs above 61 mm/L (10 mcg/mL) and peaks below 613.5 mm/L (100 mcg/mL). The pairing of blood samples with MRS scans provides a unique dataset that directly links intravenous administration of NVD to target effects in a critical CNS region and identify NAC doses and plasma levels that alleviate a key mechanism of oxidative stress and compromised CNS energetics after HIE birth. These data provide important dosing information for future clinical trial design.

We were not as successful in identifying a direct CNS target of 1,25(OH)$_2$D effect, as vitamin D's effect did not manifest as acute changes by serial MRS. However, we again demonstrate that the majority of HIE neonates are vitamin D insufficient and the standard 400IU supplemental dose of 25(OH)D is inadequate to correct this insufficiency, as we described in our previous RCT trial of hypothermia [57]. Our studies and other studies of vitamin D deficiency in HIE highlight a significant unmet need for these critically ill infants with neuro- and systemic inflammation [69,70]. Severe vitamin D deficiency is well described to impart a significant risk for worse outcomes after stroke in adult humans and animal models [71–73]. Additionally, in neonatal HIE, lower 25(OH)D concentrations in the first 48 h inversely correlated with MRI evidence of brain injury [70]. Conversely, stroke results in lower vitamin D levels through inflammation-triggered degradation of vitamin D [16,74]. Vitamin D has been shown to be neuroprotective when given primarily before and, in one study, after ischemia reperfusion injury [38,47,75,76]. 1,25(OH)$_2$D may act by regulating NMDA receptors and protecting against glutamate excitotoxicity [42,76], reducing inflammatory cytokines, modulating regulatory T cells [16,77], or chronically upregulating GSH synthetic enzymes and increasing GSH production [42–44]. While the evidence in general suggests that vitamin D deficiency appears to be detrimental for adult brain after stroke, vitamin D deficiency after global stroke in neonates may be particularly impactful, given vitamin D's important effects on plasticity, cognition and memory during brain development [16,38–40,78,79]. Based on our finding and those of others we suggest that all HIE patients have their vitamin status determined on admission and treated appropriately to achieve a serum 25(OH)D > 50 nmol/L.

Our study revealed mildly elevated ionized calcium even with q24h calcitriol dosing, which was managed by decreasing intravenous calcium or holding calcitriol doses. While we saw no clinical adverse effects from this biochemical finding, we were only able to complete an average of six out of ten days of calcitriol dosing per our safety protocol. As 25(OH)D serum concentrations had not reached sufficiency by the end of the study, increased 25(OH)D supplementation should be considered upon reaching seven days, full oral feeds, or discharge from the hospital.

We used much lower doses of NAC than employed in acetaminophen toxicity protocols or in maternal inflammation animal models of neonatal brain injury (200–300 mg/kg) [8,9,25,80,81]. Results from our trial confirm our NAC dose selection based on our preclinical studies [14,16,37,82], and demonstrate desired effects on CNS and systemic oxidative stress with NAC 25–40 mg/kg in HIE neonates without significant side effects. NAC appears to be safe in critically ill HIE neonates, just as we demonstrated in neonates of all gestational ages exposed to maternal intrauterine inflammation in our trial of NAC in maternal chorioamnionitis [83,84]. In addition to our clinical study, two other preclinical studies have demonstrated inhibition of peroxynitrite and reduction of oxidative stress with NAC 20–40 mg/kg [25,80].

We observed no significant motor, cognitive or behavioral deficits in the 24 HIE infants in pediatric follow-up, particularly impressive considering six infants had cord pH 6.5–6.8, one with no detectable heart rate until 25 min of life. Based on historical and more recent studies of hypothermia outcomes, we would expect between 8–12 infants in our cohort to have significant disability at two to four years of age [85,86]. This clinical trial extends our preclinical studies of the neuroprotective effects of NVD and demonstrates a likely neuroprotective effect in HIE neonates with severe HIE (53%) and chorioamnionitis (42%) [14,16]. We have previously shown that oxidative stress is still increasing after two hours of hypothermia in a lipopolysaccharide-HI animal model, indicating that hypothermia acts too slowly to affect this severe injury [14]. In contrast, NAC and 1,25(OH)2D rapidly corrected glutathione and suppressed glutamate in these hypothermic LPS-HI rat pups. Taken together, these preclinical and clinical studies indicate that NAC and vitamin D can mitigate overwhelming oxidative stress, treating gaps in hypothermic neuroprotection that occur both early during severe HI injury and later in the convalescent phase.

We describe a strong trend to lower [GSH] in males than in females on DOL 5, but we note that sexes were unbalanced in our cohort, with twice as many males as females, reducing our ability to accurately evaluate sex differences. However, other investigators have noted sex differences with greater oxidative stress and neural damage in males or male neurons after HI than in females, as we observed in our preclinical studies, in which low dose NAC alone did not neuroprotect males [30,87–89].

This study adds to other work demonstrating that MRS is a quantitative translational method for measuring neural injury (NAA, tCr) and antioxidant status (GSH) in HIE neonates and adults with stroke and demonstrates the unique ability to assess direct therapeutic response in a highly susceptible area of the CNS in clinical trials [15,86,90]. MRS with NAA, lactate, choline and creatine metabolite ratios are less reliable as pharmacodynamic markers, as both numerator and denominator change rapidly after stroke with their own time courses [19,90,91]. We demonstrate that serial [GSH] by MRS generates a CNS-specific, quantitative, PD time-course of treatment effect, informs effective dosing of antioxidants for a target CNS tissue/tract, and increases the likelihood of attaining the primary endpoint in subsequent clinical trials. As MRS measurements of NAA provide the best prognostic biomarker for prediction of disability early after HI injury [86,92,93], a simplified MRS protocol with a single voxel in the BG at low TE captures both GSH and NAA and would be feasible for phase II clinical trial endpoints.

This work has several limitations, chiefly the smaller sample size in some groups. We could not draw normothermia blood samples for $1,25(OH)_2D$ PK when calcitriol doses were held or were timed differently (q24h) than NAC doses, and did not coincide with the MRI scans, which had to be conducted at night. This resulted in fewer samples for $1,25(OH)_2D$ PK. We also had $1,25(OH)_2D$ serum concentrations that fluctuated and prevented PK estimate calculations and comparisons between dosing groups. We did not have MRS data immediately after HIE birth, and thus cannot quantify CNS oxidative stress in the earliest phase of injury. Our MRS scans were not optimized for lactate, but rather for GSH, and showed large coefficients of variation on LC Model spectral fitting of lactate, indicating poor confidence in isolation of lactate peaks from overlapping macromolecular and lipid peaks. We did not have funding for Bayley developmental testing and relied on other validated tests that are strongly correlated with the Bayley and are routinely performed in our high-risk clinic, as well as parent responses for two infants who had moved from the area.

5. Conclusions

In summary, in this prospective trial of NAC and $1,25(OH)_2D$ in hypothermic HIE neonates, we provide rigorous data demonstrating that CNS oxidative stress and compromised CNS energetics persist during the first week after hypoxia ischemic birth, and that NAC 25–40 mg/kg/dose and calcitriol result in rapid synthesis of GSH and increases in CNS energetics that are titrated based on the level of CNS demand from oxidative stress in the BG. With the favorable neurodevelopmental outcomes in this human trial and our preclinical animal studies, these data provide important dosing information for future clinical trial design and a strong basis for a randomized trial of NAC and $1,25(OH)_2D$ in HIE neonates.

Supplementary Materials: The following are available online at https://www.mdpi.com/article/10.3390/antiox10091344/s1, Table S1. NAC PK (25 & 40 mg/kg/dose) for individual participants; Table S2. $1,25(OH)_2D$ (0.05 mcg/kg/dose) PK for individual participants.

Author Contributions: Conceptualization, D.D.J., D.B.W. and T.R.B.; methodology, D.D.J., D.B.W., H.G.M., T.R.B., B.W.H., M.V., J.K., S.T. and M.Y.; software, H.G.M.; validation, H.G.M., T.R.B., B.W.H. and D.B.W.; formal analysis, D.D.J., B.W.H., D.B.W., H.G.M., T.R.B., M.V. and J.K.; investigation, D.D.J., C.W., B.W.H., D.B.W. and H.G.M.; resources, D.D.J., S.T., M.V. and J.K.; data curation, D.D.J., H.G.M., D.B.W., B.W.H., M.V., J.K., P.M. and M.Y.; writing—original draft preparation, D.D.J.; writing—review and editing, D.D.J., D.B.W., H.G.M., T.R.B., S.T., M.V., J.K., M.Y., C.W., B.W.H. and P.M.; visualization, D.D.J., D.B.W. and T.R.B.; supervision, D.D.J., T.R.B., M.V., S.T. and J.K.; project administration, D.D.J.; funding acquisition, D.D.J., S.T., M.V. and J.K. All authors have read and agreed to the published version of the manuscript.

Funding: This work was supported by a grant from the Medical University of South Carolina's Neuroscience Institute (Jenkins, D); NIH NINDS F31NS108623 (Moss, H); the Instituto Carlos III (FISPI14/0433), Spanish Ministry of Health, Social Services and Equality (EC11-246) (Vento, M and Kuligowski, J); and Institute of Translational Medicine & Therapeutics (ITMAT), National Institute for Health Research (NIHR) Biomedical Research Centre (BRC), Imperial College London (P60743) (Thayyil, S).

Institutional Review Board Statement: The study was conducted according to the guidelines of the Declaration of Helsinki and approved by the Medical University of South Carolina's Institutional Review Board (protocol # 31254, approval date 4 July 2014).

Informed Consent Statement: Informed consent was obtained from all subjects involved in the study.

Data Availability Statement: Individual NAC and vitamin D PK data are available in Supplemental data. Raw MRS data were generated at MUSC. Derived MRS data supporting the findings of this study are available with reasonable request.

Antioxidants **2021**, *10*, 1344

Acknowledgments: We wish to acknowledge Donnie Beason and James Purl for their dedication and expertise in performing all the MR scans.

Conflicts of Interest: The authors declare no conflict of interest. The funders had no role in the design of the study; in the collection, analyses, or interpretation of data; in the writing of the manuscript, or in the decision to publish the results.

Abbreviations

AUC = area under the curve; BG = basal ganglia; CL = clearance; $Cmax_{ss}$ = concentration maximum at steady state; $Cmin_{ss}$ = concentration minimum at steady state; DOL = day of life; ECMO = extracorporeal membrane oxygenation; FAS-FAS cell surface death receptor; GC-MS = gas chromatography-mass spectrometry; GSH = reduced glutathione; GLX = glutamate + glutamine; HI = hypoxia ischemia; HIE = hypoxic ischemic encephalopathy; HPLC = high pressure liquid chromatography; HT = hypothermia; IRB = Institutional Review Board; IVH = intraventricular hemorrhage; mIns = myoinositol; MRS = magnetic resonance spectroscopy; NAA = N-acetylaspartate + N-acetylaspartylglutamate; NAC = N-acetylcysteine; NMDA = N-methyl-D-Aspartate; NT = normothermia; NVD = NAC + vitamin D (calcitriol); PD = pharmacodynamic; PK = pharmacokinetic; PPHN = persistent pulmonary hypertension; PRESS = single point resolved spectroscopy; RIA = radioimmunoassay; ROS = reactive oxygen species; STEAM = stimulation echo acquisition mode; $t_{1/2}$ = half-life; TE = echo time; tCho = glycerophosphocholine + phosphocholine; tCr = creatine + phosphocreatine; Vd = volume of distribution.

References

1. Osredkar, D.; Sabir, H.; Falck, M.; Wood, T.; Maes, E.; Flatebo, T.; Puchades, M.; Thoresen, M. Hypothermia Does Not Reverse Cellular Responses Caused by Lipopolysaccharide in Neonatal Hypoxic-Ischaemic Brain Injury. *Dev. Neurosci.* **2015**, *37*, 390–397. [CrossRef] [PubMed]
2. Osredkar, D.; Thoresen, M.; Maes, E.; Flatebo, T.; Elstad, M.; Sabir, H. Hypothermia is not neuroprotective after infection-sensitized neonatal hypoxic-ischemic brain injury. *Resuscitation* **2013**, *85*, 567–572. [CrossRef] [PubMed]
3. Jacobs, S.E.; Berg, M.; Hunt, R.; Tarnow-Mordi, W.O.; Inder, T.E.; Davis, P.G. Cooling for newborns with hypoxic ischaemic encephalopathy. *Cochrane Database Syst. Rev.* **2013**, *126*, CD003311. [CrossRef]
4. Falck, M.; Osredkar, D.; Maes, E.; Flatebo, T.; Wood, T.R.; Sabir, H.; Thoresen, M. Hypothermic Neuronal Rescue from Infection-Sensitised Hypoxic-Ischaemic Brain Injury Is Pathogen Dependent. *Dev. Neurosci.* **2017**, *39*, 238–247. [CrossRef]
5. Sabir, H.; Scull-Brown, E.; Liu, X.; Thoresen, M. Immediate hypothermia is not neuroprotective after severe hypoxia-ischemia and is deleterious when delayed by 12 hours in neonatal rats. *Stroke* **2012**, *43*, 3364–3370. [CrossRef]
6. Barks, J.D.; Liu, Y.Q.; Shangguan, Y.; Li, J.; Pfau, J.; Silverstein, F.S. Impact of indolent inflammation on neonatal hypoxic-ischemic brain injury in mice. *Int. J. Dev. Neurosci.* **2008**, *26*, 57–65. [CrossRef]
7. Lante, F.; Meunier, J.; Guiramand, J.; De Jesus Ferreira, M.C.; Cambonie, G.; Aimar, R.; Cohen-Solal, C.; Maurice, T.; Vignes, M.; Barbanel, G. Late N-acetylcysteine treatment prevents the deficits induced in the offspring of dams exposed to an immune stress during gestation. *Hippocampus* **2008**, *18*, 602–609. [CrossRef]
8. Awad, N.; Khatib, N.; Ginsberg, Y.; Weiner, Z.; Maravi, N.; Thaler, I.; Ross, M.G.; Itsokovitz-Eldor, J.; Beloosesky, R. N-acetylcysteine (NAC) attenuates LPS-induced maternal and amniotic fluid oxidative stress and inflammatory responses in the preterm gestation. *Am. J. Obstet. Gynecol.* **2011**, *204*, 450.e15–450.e20. [CrossRef] [PubMed]
9. Wang, X.; Svedin, P.; Nie, C.; Lapatto, R.; Zhu, C.; Gustavsson, M.; Sandberg, M.; Karlsson, J.O.; Romero, R.; Hagberg, H.; et al. N-acetylcysteine reduces lipopolysaccharide-sensitized hypoxic-ischemic brain injury. *Ann. Neurol.* **2007**, *61*, 263–271. [CrossRef]
10. Liu, J.Q.; Lee, T.F.; Chen, C.; Bagim, D.L.; Cheung, P.Y. N-acetylcysteine improves hemodynamics and reduces oxidative stress in the brains of newborn piglets with hypoxia-reoxygenation injury. *J. Neurotrauma* **2010**, *27*, 1865–1873. [CrossRef]
11. Vento, M.; Asensi, M.; Sastre, J.; Lloret, A.; Garcia-Sala, F.; Vina, J. Oxidative stress in asphyxiated term infants resuscitated with 100% oxygen. *J. Pediatr.* **2003**, *142*, 240–246. [CrossRef] [PubMed]
12. Armstrong, J.S.; Steinauer, K.K.; Hornung, B.; Irish, J.M.; Lecane, P.; Birrell, G.W.; Peehl, D.M.; Knox, S.J. Role of glutathione depletion and reactive oxygen species generation in apoptotic signaling in a human B lymphoma cell line. *Cell Death Differ.* **2002**, *9*, 252–263. [CrossRef]
13. Rae, C.D.; Williams, S.R. Glutathione in the human brain: Review of its roles and measurement by magnetic resonance spectroscopy. *Anal. Biochem.* **2017**, *529*, 127–143. [CrossRef] [PubMed]
14. Adams, L.E.; Moss, H.G.; Lowe, D.W.; Brown, T.; Wiest, D.B.; Hollis, B.W.; Singh, I.; Jenkins, D.D. NAC and Vitamin D Restore CNS Glutathione in Endotoxin-Sensitized Neonatal Hypoxic-Ischemic Rats. *Antioxidants* **2021**, *10*, 489. [CrossRef] [PubMed]

15. Moss, H.G.; Brown, T.R.; Wiest, D.B.; Jenkins, D.D. N-Acetylcysteine rapidly replenishes central nervous system glutathione measured via magnetic resonance spectroscopy in human neonates with hypoxic-ischemic encephalopathy. *J. Cereb. Blood Flow Metab.* **2018**, *38*, 950–958. [CrossRef]
16. Lowe, D.W.; Fraser, J.L.; Rollins, L.G.; Bentzley, J.; Nie, X.; Martin, R.; Singh, I.; Jenkins, D. Vitamin D improves functional outcomes in neonatal hypoxic ischemic male rats treated with N-acetylcysteine and hypothermia. *Neuropharmacology* **2017**, *123*, 186–200. [CrossRef]
17. Selakovic, V.; Korenic, A.; Radenovic, L. Spatial and temporal patterns of oxidative stress in the brain of gerbils submitted to different duration of global cerebral ischemia. *Int. J. Dev. Neurosci.* **2011**, *29*, 645–654. [CrossRef]
18. Sanchez-Illana, A.; Thayyil, S.; Montaldo, P.; Jenkins, D.; Quintas, G.; Oger, C.; Galano, J.M.; Vigor, C.; Durand, T.; Vento, M.; et al. Novel free-radical mediated lipid peroxidation biomarkers in newborn plasma. *Anal. Chim. Acta* **2017**, *996*, 88–97. [CrossRef]
19. Munoz Maniega, S.; Cvoro, V.; Chappell, F.M.; Armitage, P.A.; Marshall, I.; Bastin, M.E.; Wardlaw, J.M. Changes in NAA and lactate following ischemic stroke: A serial MR spectroscopic imaging study. *Neurology* **2008**, *71*, 1993–1999. [CrossRef]
20. Franco, R.; Bortner, C.D.; Schmitz, I.; Cidlowski, J.A. Glutathione depletion regulates both extrinsic and intrinsic apoptotic signaling cascades independent from multidrug resistance protein 1. *Apoptosis* **2014**, *19*, 117–134. [CrossRef]
21. Ibi, M.; Sawada, H.; Kume, T.; Katsuki, H.; Kaneko, S.; Shimohama, S.; Akaike, A. Depletion of intracellular glutathione increases susceptibility to nitric oxide in mesencephalic dopaminergic neurons. *J. Neurochem.* **1999**, *73*, 1696–1703. [CrossRef]
22. Quintana-Cabrera, R.; Bolanos, J.P. Glutathione and gamma-glutamylcysteine in the antioxidant and survival functions of mitochondria. *Biochem. Soc. Trans.* **2013**, *41*, 106–110. [CrossRef]
23. Ceccon, M.; Giusti, P.; Facci, L.; Borin, G.; Imbesi, M.; Floreani, M.; Skaper, S.D. Intracellular glutathione levels determine cerebellar granule neuron sensitivity to excitotoxic injury by kainic acid. *Brain Res.* **2000**, *862*, 83–89. [CrossRef]
24. Shih, A.Y.; Johnson, D.A.; Wong, G.; Kraft, A.D.; Jiang, L.; Erb, H.; Johnson, J.A.; Murphy, T.H. Coordinate regulation of glutathione biosynthesis and release by Nrf2-expressing glia potently protects neurons from oxidative stress. *J. Neurosci.* **2003**, *23*, 3394–3406. [CrossRef]
25. Cuzzocrea, S.; Mazzon, E.; Costantino, G.; Serraino, I.; Dugo, L.; Calabro, G.; Cucinotta, G.; De Sarro, A.; Caputi, A.P. Beneficial effects of n-acetylcysteine on ischaemic brain injury. *Br. J. Pharmacol.* **2000**, *130*, 1219–1226. [CrossRef]
26. Victor, V.M.; Rocha, M.; De la Fuente, M. N-acetylcysteine protects mice from lethal endotoxemia by regulating the redox state of immune cells. *Free Radic. Res.* **2003**, *37*, 919–929. [CrossRef]
27. Katz, M.; Won, S.J.; Park, Y.; Orr, A.; Jones, D.P.; Swanson, R.A.; Glass, G.A. Cerebrospinal fluid concentrations of N-acetylcysteine after oral administration in Parkinson's disease. *Parkinsonism Relat. Disord.* **2015**, *21*, 500–503. [CrossRef]
28. Neuwelt, E.A.; Pagel, M.A.; Hasler, B.P.; Deloughery, T.G.; Muldoon, L.L. Therapeutic efficacy of aortic administration of N-acetylcysteine as a chemoprotectant against bone marrow toxicity after intracarotid administration of alkylators, with or without glutathione depletion in a rat model. *Cancer Res.* **2001**, *61*, 7868–7874. [PubMed]
29. Holmay, M.J.; Terpstra, M.; Coles, L.D.; Mishra, U.; Ahlskog, M.; Oz, G.; Cloyd, J.C.; Tuite, P.J. N-Acetylcysteine boosts brain and blood glutathione in Gaucher and Parkinson diseases. *Clin. Neuropharmacol.* **2013**, *36*, 103–106. [CrossRef]
30. Du, L.; Empey, P.E.; Ji, J.; Chao, H.; Kochanek, P.M.; Bayir, H.; Clark, R.S. Probenecid and N-Acetylcysteine Prevent Loss of Intracellular Glutathione and Inhibit Neuronal Death after Mechanical Stretch Injury In Vitro. *J. Neurotrauma* **2016**, *33*, 1913–1917. [CrossRef] [PubMed]
31. Choy, K.H.; Dean, O.; Berk, M.; Bush, A.I.; van den Buuse, M. Effects of N-acetyl-cysteine treatment on glutathione depletion and a short-term spatial memory deficit in 2-cyclohexene-1-one-treated rats. *Eur. J. Pharmacol.* **2010**, *649*, 224–228. [CrossRef]
32. Mahmoud, S.M.; Abdel Moneim, A.E.; Qayed, M.M.; El-Yamany, N.A. Potential role of N-acetylcysteine on chlorpyrifos-induced neurotoxicity in rats. *Environ. Sci. Pollut. Res. Int.* **2019**, *26*, 20731–20741. [CrossRef] [PubMed]
33. Farr, S.A.; Poon, H.F.; Dogrukol-Ak, D.; Drake, J.; Banks, W.A.; Eyerman, E.; Butterfield, D.A.; Morley, J.E. The antioxidants alpha-lipoic acid and N-acetylcysteine reverse memory impairment and brain oxidative stress in aged SAMP8 mice. *J. Neurochem.* **2003**, *84*, 1173–1183. [CrossRef]
34. Khan, M.; Sekhon, B.; Jatana, M.; Giri, S.; Gilg, A.G.; Sekhon, C.; Singh, I.; Singh, A.K. Administration of N-acetylcysteine after focal cerebral ischemia protects brain and reduces inflammation in a rat model of experimental stroke. *J. Neurosci. Res.* **2004**, *76*, 519–527. [CrossRef]
35. Lee, T.F.; Jantzie, L.L.; Todd, K.G.; Cheung, P.Y. Postresuscitation N-acetylcysteine treatment reduces cerebral hydrogen peroxide in the hypoxic piglet brain. *Intensive Care Med.* **2008**, *34*, 190–197. [CrossRef]
36. Park, D.; Shin, K.; Choi, E.K.; Choi, Y.; Jang, J.Y.; Kim, J.; Jeong, H.S.; Lee, W.; Lee, Y.B.; Kim, S.U.; et al. Protective effects of N-acetyl-L-cysteine in human oligodendrocyte progenitor cells and restoration of motor function in neonatal rats with hypoxic-ischemic encephalopathy. *Evid. Based Complement. Alternat. Med.* **2015**, *2015*, 764251. [CrossRef]
37. Nie, X.; Lowe, D.W.; Rollins, L.G.; Bentzley, J.; Fraser, J.L.; Martin, R.; Singh, I.; Jenkins, D. Sex-specific effects of N-Acetylcysteine in neonatal rats treated with hypothermia after severe hypoxia-ischemia. *Neurosci. Res.* **2016**, *108*, 24–33. [CrossRef]
38. Gomez-Pinedo, U.; Cuevas, J.A.; Benito-Martin, M.S.; Moreno-Jimenez, L.; Esteban-Garcia, N.; Torre-Fuentes, L.; Matias-Guiu, J.A.; Pytel, V.; Montero, P.; Matias-Guiu, J. Vitamin D increases remyelination by promoting oligodendrocyte lineage differentiation. *Brain Behav.* **2020**, *10*, e01498. [CrossRef]

39. Eyles, D.; Almeras, L.; Benech, P.; Patatian, A.; Mackay-Sim, A.; McGrath, J.; Feron, F. Developmental vitamin D deficiency alters the expression of genes encoding mitochondrial, cytoskeletal and synaptic proteins in the adult rat brain. *J. Steroid Biochem. Mol. Biol.* **2007**, *103*, 538–545. [CrossRef]

40. Cui, X.; McGrath, J.J.; Burne, T.H.; Mackay-Sim, A.; Eyles, D.W. Maternal vitamin D depletion alters neurogenesis in the developing rat brain. *Int. J. Dev. Neurosci.* **2007**, *25*, 227–232. [CrossRef]

41. George, N.; Kumar, T.P.; Antony, S.; Jayanarayanan, S.; Paulose, C.S. Effect of vitamin D3 in reducing metabolic and oxidative stress in the liver of streptozotocin-induced diabetic rats. *Br. J. Nutr.* **2012**, *108*, 1410–1418. [CrossRef]

42. Ibi, M.; Sawada, H.; Nakanishi, M.; Kume, T.; Katsuki, H.; Kaneko, S.; Shimohama, S.; Akaike, A. Protective effects of 1 alpha,25-(OH)(2)D(3) against the neurotoxicity of glutamate and reactive oxygen species in mesencephalic culture. *Neuropharmacology* **2001**, *40*, 761–771. [CrossRef]

43. Jain, S.K.; Micinski, D. Vitamin D upregulates glutamate cysteine ligase and glutathione reductase, and GSH formation, and decreases ROS and MCP-1 and IL-8 secretion in high-glucose exposed U937 monocytes. *Biochem. Biophys. Res. Commun.* **2013**, *437*, 7–11. [CrossRef]

44. Garcion, E.; Sindji, L.; Leblondel, G.; Brachet, P.; Darcy, F. 1.25-dihydroxyvitamin D3 regulates the synthesis of gamma-glutamyl transpeptidase and glutathione levels in rat primary astrocytes. *J. Neurochem.* **1999**, *73*, 859–866. [CrossRef]

45. Jain, S.K.; Micinski, D.; Huning, L.; Kahlon, G.; Bass, P.F.; Levine, S.N. Vitamin D and L-cysteine levels correlate positively with GSH and negatively with insulin resistance levels in the blood of type 2 diabetic patients. *Eur J. Clin. Nutr.* **2014**, *68*, 1148–1153. [CrossRef] [PubMed]

46. Kajta, M.; Makarewicz, D.; Zieminska, E.; Jantas, D.; Domin, H.; Lason, W.; Kutner, A.; Lazarewicz, J.W. Neuroprotection by co-treatment and post-treating with calcitriol following the ischemic and excitotoxic insult in vivo and in vitro. *Neurochem. Int.* **2009**, *55*, 265–274. [CrossRef]

47. Wang, Y.; Chiang, Y.H.; Su, T.P.; Hayashi, T.; Morales, M.; Hoffer, B.J.; Lin, S.Z. Vitamin D(3) attenuates cortical infarction induced by middle cerebral arterial ligation in rats. *Neuropharmacology* **2000**, *39*, 873–880. [CrossRef]

48. Wijtenburg, S.A.; Gaston, F.E.; Spieker, E.A.; Korenic, S.A.; Kochunov, P.; Hong, L.E.; Rowland, L.M. Reproducibility of phase rotation STEAM at 3T: Focus on glutathione. *Magn. Reson. Med.* **2014**, *72*, 603–609. [CrossRef]

49. Wijtenburg, S.A.; Near, J.; Korenic, S.A.; Gaston, F.E.; Chen, H.; Mikkelsen, M.; Chen, S.; Kochunov, P.; Hong, L.E.; Rowland, L.M. Comparing the reproducibility of commonly used magnetic resonance spectroscopy techniques to quantify cerebral glutathione. *J. Magn. Reson. Imaging* **2019**, *49*, 176–183. [CrossRef] [PubMed]

50. Tkac, I.; Oz, G.; Adriany, G.; Ugurbil, K.; Gruetter, R. In vivo 1H NMR spectroscopy of the human brain at high magnetic fields: Metabolite quantification at 4T vs. 7T. *Magn. Reson. Med.* **2009**, *62*, 868–879. [CrossRef] [PubMed]

51. Coles, L.D.; Tuite, P.J.; Oz, G.; Mishra, U.R.; Kartha, R.V.; Sullivan, K.M.; Cloyd, J.C.; Terpstra, M. Repeated-Dose Oral N-Acetylcysteine in Parkinson's Disease: Pharmacokinetics and Effect on Brain Glutathione and Oxidative Stress. *J. Clin. Pharmacol.* **2018**, *58*, 158–167. [CrossRef]

52. Kreis, R.; Hofmann, L.; Kuhlmann, B.; Boesch, C.; Bossi, E.; Huppi, P.S. Brain metabolite composition during early human brain development as measured by quantitative in vivo 1H magnetic resonance spectroscopy. *Magn. Reson. Med.* **2002**, *48*, 949–958. [CrossRef] [PubMed]

53. Sarnat, H.B.; Sarnat, M.S. Neonatal encephalopathy following fetal distress. A clinical and electroencephalographic study. *Arch. Neurol.* **1976**, *33*, 696–705. [CrossRef]

54. Sarnat, H.B.; Flores-Sarnat, L.; Fajardo, C.; Leijser, L.M.; Wusthoff, C.; Mohammad, K. Sarnat Grading Scale for Neonatal Encephalopathy after 45 Years: An Update Proposal. *Pediatr. Neurol.* **2020**, *113*, 75–79. [CrossRef] [PubMed]

55. Eicher, D.J.; Wagner, C.L.; Katikaneni, L.P.; Hulsey, T.C.; Bass, W.T.; Kaufman, D.A.; Horgan, M.J.; Languani, S.; Bhatia, J.J.; Givelichian, L.M.; et al. Moderate hypothermia in neonatal encephalopathy: Safety outcomes. *Pediatr. Neurol.* **2005**, *32*, 18–24. [CrossRef] [PubMed]

56. Ercal, N.; Oztezcan, S.; Hammond, T.C.; Matthews, R.H.; Spitz, D.R. High-performance liquid chromatography assay for N-acetylcysteine in biological samples following derivatization with N-(1-pyrenyl)maleimide. *J. Chromatogr. B Biomed. Appl.* **1996**, *685*, 329–334. [CrossRef]

57. Lowe, D.W.; Hollis, B.W.; Wagner, C.L.; Bass, T.; Kaufman, D.A.; Horgan, M.J.; Givelichian, L.M.; Sankaran, K.; Yager, J.Y.; Katikaneni, L.D.; et al. Vitamin D Insufficiency in Neonatal Hypoxic-Ischemic Encephalopathy. *Pediatr. Res.* **2017**, *82*, 55–62. [CrossRef]

58. Gill, K.; Osiovich, A.; Synnes, A.; Agnew, J.A.; Grunau, R.E.; Miller, S.P.; Zwicker, J.G. Concurrent Validity of the Bayley-III and the Peabody Developmental Motor Scales-2 at 18 Months. *Phys. Occup. Ther. Pediatr.* **2019**, *39*, 514–524. [CrossRef]

59. Hoon, A.H., Jr.; Pulsifer, M.B.; Gopalan, R.; Palmer, F.B.; Capute, A.J. Clinical Adaptive Test/Clinical Linguistic Auditory Milestone Scale in early cognitive assessment. *J. Pediatr.* **1993**, *123*, S1–S8. [CrossRef]

60. Zamora, C.; Tekes, A.; Alqahtani, E.; Kalayci, O.T.; Northington, F.; Huisman, T.A. Variability of resistive indices in the anterior cerebral artery during fontanel compression in preterm and term neonates measured by transcranial duplex sonography. *J. Perinatol. Off. J. Calif. Perinat. Assoc.* **2014**, *34*, 306–310. [CrossRef]

61. Allison, J.W.; Faddis, L.A.; Kinder, D.L.; Roberson, P.K.; Glasier, C.M.; Seibert, J.J. Intracranial resistive index (RI) values in normal term infants during the first day of life. *Pediatr. Radiol.* **2000**, *30*, 618–620. [CrossRef] [PubMed]

62. Zella, L.A.; Shevde, N.K.; Hollis, B.W.; Cooke, N.E.; Pike, J.W. Vitamin D-binding protein influences total circulating levels of 1,25-dihydroxyvitamin D3 but does not directly modulate the bioactive levels of the hormone in vivo. *Endocrinology* **2008**, *149*, 3656–3667. [CrossRef] [PubMed]

63. Zhang, H.; Limphong, P.; Pieper, J.; Liu, Q.; Rodesch, C.K.; Christians, E.; Benjamin, I.J. Glutathione-dependent reductive stress triggers mitochondrial oxidation and cytotoxicity. *FASEB J.* **2012**, *26*, 1442–1451. [CrossRef] [PubMed]

64. Andres, R.H.; Ducray, A.D.; Schlattner, U.; Wallimann, T.; Widmer, H.R. Functions and effects of creatine in the central nervous system. *Brain Res. Bull.* **2008**, *76*, 329–343. [CrossRef] [PubMed]

65. Lu, S.C. Glutathione synthesis. *Biochim Biophys. Acta* **2013**, *1830*, 3143–3153. [CrossRef]

66. Cuyamendous, C.; de la Torre, A.; Lee, Y.Y.; Leung, K.S.; Guy, A.; Bultel-Ponce, V.; Galano, J.M.; Lee, J.C.; Oger, C.; Durand, T. The novelty of phytofurans, isofurans, dihomo-isofurans and neurofurans: Discovery, synthesis and potential application. *Biochimie* **2016**, *130*, 49–62. [CrossRef]

67. Roka, A.; Melinda, K.T.; Vasarhelyi, B.; Machay, T.; Azzopardi, D.; Szabo, M. Elevated morphine concentrations in neonates treated with morphine and prolonged hypothermia for hypoxic ischemic encephalopathy. *Pediatrics* **2008**, *121*, e844–e849. [CrossRef] [PubMed]

68. Filippi, L.; la Marca, G.; Cavallaro, G.; Fiorini, P.; Favelli, F.; Malvagia, S.; Donzelli, G.; Guerrini, R. Phenobarbital for neonatal seizures in hypoxic ischemic encephalopathy: A pharmacokinetic study during whole body hypothermia. *Epilepsia* **2011**, *52*, 794–801. [CrossRef]

69. Mutlu, M.; Sariaydin, M.; Aslan, Y.; Kader, S.; Dereci, S.; Kart, C.; Yaman, S.O.; Kural, B. Status of vitamin D, antioxidant enzymes, and antioxidant substances in neonates with neonatal hypoxic-ischemic encephalopathy. *J. Matern.-Fetal Neonatal Med.* **2016**, *29*, 2259–2263. [CrossRef]

70. McGinn, E.A.; Powers, A.; Galas, M.; Lyden, E.; Peeples, E.S. Neonatal Vitamin D Status Is Associated with the Severity of Brain Injury in Neonatal Hypoxic-Ischemic Encephalopathy: A Pilot Study. *Neuropediatrics* **2020**, *51*, 251–258. [CrossRef] [PubMed]

71. Balden, R.; Selvamani, A.; Sohrabji, F. Vitamin D deficiency exacerbates experimental stroke injury and dysregulates ischemia-induced inflammation in adult rats. *Endocrinology* **2012**, *153*, 2420–2435. [CrossRef]

72. Park, K.Y.; Chung, P.W.; Kim, Y.B.; Moon, H.S.; Suh, B.C.; Won, Y.S.; Kim, J.M.; Youn, Y.C.; Kwon, O.S. Serum Vitamin D Status as a Predictor of Prognosis in Patients with Acute Ischemic Stroke. *Cerebrovasc. Dis.* **2015**, *40*, 73–80. [CrossRef]

73. Turetsky, A.; Goddeau, R.P., Jr.; Henninger, N. Low Serum Vitamin D Is Independently Associated with Larger Lesion Volumes after Ischemic Stroke. *J. Stroke Cerebrovasc. Dis. Off. J. Natl. Stroke Assoc.* **2015**, *24*, 1555–1563. [CrossRef] [PubMed]

74. Berghout, B.P.; Fani, L.; Heshmatollah, A.; Koudstaal, P.J.; Ikram, M.A.; Zillikens, M.C.; Ikram, M.K. Vitamin D Status and Risk of Stroke: The Rotterdam Study. *Stroke* **2019**, *50*, 2293–2298. [CrossRef] [PubMed]

75. Brewer, L.D.; Thibault, V.; Chen, K.C.; Langub, M.C.; Landfield, P.W.; Porter, N.M. Vitamin D hormone confers neuroprotection in parallel with downregulation of L-type calcium channel expression in hippocampal neurons. *J. Neurosci.* **2001**, *21*, 98–108. [CrossRef] [PubMed]

76. Fu, J.; Xue, R.; Gu, J.; Xiao, Y.; Zhong, H.; Pan, X.; Ran, R. Neuroprotective effect of calcitriol on ischemic/reperfusion injury through the NR3A/CREB pathways in the rat hippocampus. *Mol. Med. Rep.* **2013**, *8*, 1708–1714. [CrossRef] [PubMed]

77. Evans, M.A.; Kim, H.A.; Ling, Y.H.; Uong, S.; Vinh, A.; De Silva, T.M.; Arumugam, T.V.; Clarkson, A.N.; Zosky, G.R.; Drummond, G.R.; et al. Vitamin D3 Supplementation Reduces Subsequent Brain Injury and Inflammation Associated with Ischemic Stroke. *Neuromol. Med.* **2018**, *20*, 147–159. [CrossRef]

78. Becker, A.; Eyles, D.W.; McGrath, J.J.; Grecksch, G. Transient prenatal vitamin D deficiency is associated with subtle alterations in learning and memory functions in adult rats. *Behav. Brain Res.* **2005**, *161*, 306–312. [CrossRef]

79. Groves, N.J.; McGrath, J.J.; Burne, T.H. Vitamin D as a neurosteroid affecting the developing and adult brain. *Annu Rev. Nutr.* **2014**, *34*, 117–141. [CrossRef]

80. Cuzzocrea, S.; Costantino, G.; Caputi, A.P. Protective effect of N-acetylcysteine on cellular energy depletion in a non-septic shock model induced by zymosan in the rat. *Shock* **1999**, *11*, 143–148. [CrossRef]

81. Sharabi, H.; Khatib, N.; Ginsberg, Y.; Weiner, Z.; Ross, M.G.; Tamar, B.K.; Efrat, S.; Mordechai, H.; Beloosesky, R. Therapeutic N-Acetyl-Cysteine (Nac) Following Initiation of Maternal Inflammation Attenuates Long-Term Offspring Cerebral Injury, as Evident in Magnetic Resonance Imaging (MRI). *Neuroscience* **2019**, *403*, 118–124. [CrossRef]

82. Jatana, M.; Singh, I.; Singh, A.K.; Jenkins, D. Combination of systemic hypothermia and N-acetylcysteine attenuates hypoxic-ischemic brain injury in neonatal rats. *Pediatr. Res.* **2006**, *59*, 684–689. [CrossRef]

83. Jenkins, D.D.; Wiest, D.B.; Mulvihill, D.M.; Hlavacek, A.M.; Majstoravich, S.J.; Brown, T.R.; Taylor, J.J.; Buckley, J.R.; Turner, R.P.; Rollins, L.G.; et al. Fetal and Neonatal Effects of N-Acetylcysteine When Used for Neuroprotection in Maternal Chorioamnionitis. *J. Pediatr.* **2016**, *168*, 67–76. [CrossRef]

84. Wiest, D.B.; Chang, E.; Fanning, D.; Garner, S.; Cox, T.; Jenkins, D.D. Antenatal pharmacokinetics and placental transfer of N-acetylcysteine in chorioamnionitis for fetal neuroprotection. *J. Pediatr.* **2014**, *165*, 672–677. [CrossRef] [PubMed]

85. Edwards, A.D.; Brocklehurst, P.; Gunn, A.J.; Halliday, H.; Juszczak, E.; Levene, M.; Strohm, B.; Thoresen, M.; Whitelaw, A.; Azzopardi, D. Neurological outcomes at 18 months of age after moderate hypothermia for perinatal hypoxic ischaemic encephalopathy: Synthesis and meta-analysis of trial data. *BMJ* **2010**, *340*, c363. [CrossRef]

86. Lally, P.J.; Montaldo, P.; Oliveira, V.; Soe, A.; Swamy, R.; Bassett, P.; Mendoza, J.; Atreja, G.; Kariholu, U.; Pattnayak, S.; et al. Magnetic resonance spectroscopy assessment of brain injury after moderate hypothermia in neonatal encephalopathy: A prospective multicentre cohort study. *Lancet Neurol.* **2019**, *18*, 35–45. [CrossRef]
87. Demarest, T.G.; Schuh, R.A.; Waddell, J.; McKenna, M.C.; Fiskum, G. Sex-dependent mitochondrial respiratory impairment and oxidative stress in a rat model of neonatal hypoxic-ischemic encephalopathy. *J. Neurochem.* **2016**, *137*, 714–729. [CrossRef]
88. Demarest, T.G.; Schuh, R.A.; Waite, E.L.; Waddell, J.; McKenna, M.C.; Fiskum, G. Sex dependent alterations in mitochondrial electron transport chain proteins following neonatal rat cerebral hypoxic-ischemia. *J. Bioenerg. Biomembr.* **2016**, *48*, 591–598. [CrossRef] [PubMed]
89. Nijboer, C.H.; Groenendaal, F.; Kavelaars, A.; Hagberg, H.H.; van Bel, F.; Heijnen, C.J. Gender-specific neuroprotection by 2-iminobiotin after hypoxia-ischemia in the neonatal rat via a nitric oxide independent pathway. *J. Cereb. Blood Flow Metab.* **2007**, *27*, 282–292. [CrossRef]
90. Gideon, P.; Henriksen, O.; Sperling, B.; Christiansen, P.; Olsen, T.S.; Jorgensen, H.S.; Arlien-Soborg, P. Early time course of N-acetylaspartate, creatine and phosphocreatine, and compounds containing choline in the brain after acute stroke. A proton magnetic resonance spectroscopy study. *Stroke* **1992**, *23*, 1566–1572. [CrossRef]
91. Munoz Maniega, S.; Cvoro, V.; Armitage, P.A.; Marshall, I.; Bastin, M.E.; Wardlaw, J.M. Choline and creatine are not reliable denominators for calculating metabolite ratios in acute ischemic stroke. *Stroke* **2008**, *39*, 2467–2469. [CrossRef] [PubMed]
92. Robertson, N.J.; Thayyil, S.; Cady, E.B.; Raivich, G. Magnetic resonance spectroscopy biomarkers in term perinatal asphyxial encephalopathy: From neuropathological correlates to future clinical applications. *Curr Pediatr. Rev.* **2014**, *10*, 37–47. [CrossRef] [PubMed]
93. Thayyil, S.; Chandrasekaran, M.; Taylor, A.; Bainbridge, A.; Cady, E.B.; Chong, W.K.; Murad, S.; Omar, R.Z.; Robertson, N.J. Cerebral magnetic resonance biomarkers in neonatal encephalopathy: A meta-analysis. *Pediatrics* **2010**, *125*, e382–e395. [CrossRef] [PubMed]

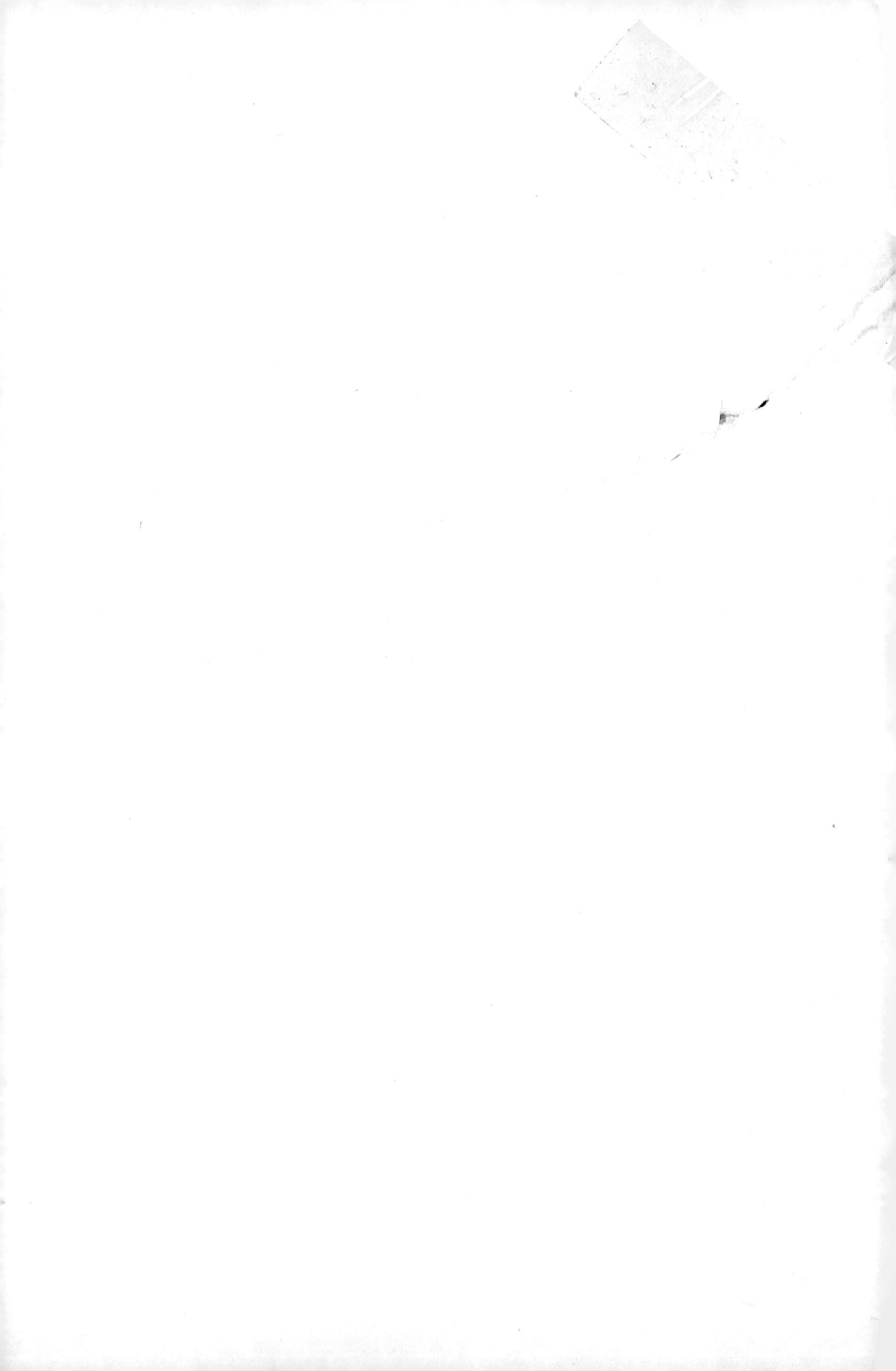

MDPI

St. Alban-Anlage 66

4052 Basel

Switzerland

Tel. +41 61 683 77 34

Fax +41 61 302 89 18

www.mdpi.com

Antioxidants Editorial Office

E-mail: antioxidants@mdpi.com

www.mdpi.com/journal/antioxidants